Springer Water

The book series Springer Water comprises a broad portfolio of multi- and interdisciplinary scientific books, aiming at researchers, students, and everyone interested in water-related science. The series includes peer-reviewed monographs, edited volumes, textbooks, and conference proceedings. Its volumes combine all kinds of water-related research areas, such as: the movement, distribution and quality of freshwater; water resources; the quality and pollution of water and its influence on health; the water industry including drinking water, wastewater, and desalination services and technologies; water history; as well as water management and the governmental, political, developmental, and ethical aspects of water.

Hamdi A. Zurqani
Editor

Water Resources of Libya

Challenges and Management

Springer

Editor
Hamdi A. Zurqani
University of Arkansas Agricultural Experiment Station, Arkansas Forest Resources Center
College of Forestry, Agriculture, and Natural Resources, University of Arkansas
Monticello, AR, USA

ISSN 2364-6934 ISSN 2364-8198 (electronic)
Springer Water
ISBN 978-3-031-80919-4 ISBN 978-3-031-80920-0 (eBook)
https://doi.org/10.1007/978-3-031-80920-0

This Springer imprint is published by the registered company Springer Nature Switzerland AG
The registered company address is: Gewerbestrasse 11, 6330 Cham, Switzerland

Preface

Water is a natural resource essential for life on Earth. It has been used for various purposes, including drinking, sanitation, irrigation, and power production. The book deals with current water resource problems in Libya and their management strategies and policies. It presents a comprehensive overview of the hydrology in Libya, including water availability and distribution. The book also sheds light on the most critical issues related to water quantity and quality in the region. It further explores ways of protecting the water resources in the country, with a particular focus on safeguarding these precious resources for future generations. In addition, it illustrates some practical examples that show how different approaches have been applied to address water problems in this country through case studies from Libya's experience concerning successful water management solutions. Moreover, the book offers a compilation of unpublished research that adds new knowledge about water management while addressing other topical issues relevant to Libyan Water Resources. This makes it an essential reference guide for experts, researchers, and professionals who work in fields such as hydrology, water management, and environmental science. The first chapter introduces the book Introduction to the *Water Resources of Libya: Challenges and Management* and summarizes all the chapters presented in this book. The second chapter gives an overview of the Great Man-Made River Project (GMMRP), including its current status and future prospects. The third chapter focuses on groundwater as an important resource of urban water supply in Libya, with each person having access to less than 10% of the global average. The fourth chapter discusses the seasonal surface water "rainwater harvesting" in Libya as sustainable solutions for water scarcity in the country, where water resources are limited due to low rainfall and reliance on finite sources. The fifth chapter discusses agricultural drainage in Libya from the 1950s to the present day and reviews the development of drainage systems, analyzes the existing systems, and proposes solutions for improvement. Furthermore, the sixth chapter reviews and assesses irrigation development in Libya from the 1960s to the 2010s and highlights the challenges the country faces in agricultural irrigation and proposes solutions to improve it. Moreover, the seventh chapter explores the challenges and opportunities related to irrigation in Libya and highlights the importance of sustainable water management practices, such as drip irrigation and rainwater harvesting, to conserve water resources. The eighth chapter discusses the causes, impacts, and required actions to reduce flood risk in Libya and provides recommendations for improving flood prevention and management strategies in the country. The ninth chapter examines the relationship between climate patterns and water supply and demand in Libya over 60 years, from 1961 to 2020, and provides valuable insights into climate change trends in the country. The final chapter summarizes and highlights the main characteristic issues of water resources in Libya, as well as the most important conclusions and recommendations from all chapters to provide perspective and guidance for students, teachers, researchers, decision-makers, etc. It also includes summarized information to provide readers with an in-depth understanding of the complexities and potential solutions to ensure appropriate management of water resources in the region. This book has a valuable contribution to many distinguished scientists from various universities and institutions in and outside of Libya.

Monticello, AR, USA
January 2025

Dr. Hamdi A. Zurqani

Acknowledgments I would like to acknowledge Dr. Margaret Deignan, senior series editor in chief of Springer Nature, and all the staff of Springer that kindly assisted us in our multiple queries during the elaboration of this book. I would also like to thank all authors who contributed to this book and whose cooperation made my task as an editor a pleasure.

About This book

This book on the water resources of Libya provides an up-to-date review of critical aspects concerning the water resources in the country, including hydrology, water quality, water quantity, and water resource protection. It outlines a comprehensive overview of the challenges facing Libyan water resources, including qualitative and quantitative dimensions of water resource management. The book provides insights into different water management strategies that can effectively reduce water waste, promote sustainable water usage, and optimize water quality and quantity in the country. It also suggests some effective water management strategies, including investing in innovative technology, reducing water waste, and taking eco-conscious measures to conserve water. Several case studies are included to demonstrate practical solutions and successful practices in water management within the Libyan context. These case studies serve as valuable learning tools, showing how theoretical knowledge can be applied to solve real-world water management issues in Libya. Moreover, the book is a treasure trove of unpublished research, offering insights into various topical issues related to water management in Libya. This book is essential for policymakers, researchers, practitioners, and students involved in hydrology, environmental science, and water resource management. It provides a deep dive into the complexities of managing water resources in Libya, where water is a precious but scarce commodity. It also highlights the need for innovative, sustainable, and community-focused approaches to water management.

Contents

Editor and Contributors

About the Editor

Dr. Hamdi A. Zurqani is an Assistant Professor of Geospatial Science in Natural Resource Management and Conservation at the University of Arkansas Agricultural Experiment Station, Arkansas Forest Resources Center, University of Arkansas at Monticello, USA. With a strong academic foundation and a passion for innovative technologies, he is also an FAA (i.e., Pt107) Certified sUAS/Drones Pilot, leveraging this expertise to enhance his proficiency in remote sensing, GIS, and geospatial data analysis, which he applies to advance his research and teaching in environmental monitoring and analysis. Dr. Zurqani is a recognized expert as a result of his internationally acclaimed work in the areas of environmental information science, remote sensing, geospatial analysis, land evaluation, sustainability, pedology, and soil science education. He has conducted research across the world, including the United States of America and Africa, and has served as PI, co-PI, or co-investigator on several grants-funded research projects. In addition to his research and teaching responsibilities, Dr. Zurqani is actively involved in institutional service at the University of Arkansas at Monticello. He serves on the UAM Campus-wide Artificial Intelligence (AI) Review Committee, the UAM Faculty Assembly Student Engagement Committee, and the UAM Library Committee.

Professionally, Dr. Zurqani contributes to the broader academic community through a variety of leadership and collaborative roles. He is an active member of the Steering Committee of the Institute of Electrical and Electronics Engineers (IEEE) Mediterranean and Middle-East Geoscience and Remote Sensing Symposium (M2GARSS) and serves as an Ambassador for the Geoscience and Remote Sensing Society (GRSS) Chapter of the IEEE in Libya, facilitating communication, collaboration, and knowledge-sharing within the geospatial and remote sensing communities. In recognition of his expertise and substantial contributions to the field, Dr. Zurqani was invited in 2024 to serve on the Technical Program Committee for the IEEE Asia-Pacific Conference on Geoscience, Electronics, and Remote Sensing Technology (AGERS). In 2025, he also joined the IEEE GRSS Technical Committee on Remote Sensing Environment, Analysis, and Climate Technologies (REACT), a platform for scientists and engineers to exchange ideas and knowledge on environmental issues and the impacts of climate change, emphasizing support for sustainable development goals (SDGs) through remote sensing technologies. Beyond these roles, Dr. Zurqani holds the position of UAM Co-Principal Investigator at the Southeast Climate Adaptation Science Center, where he is involved in research initiatives focused on climate adaptation and resilience in the southeastern United States and its surrounding regions.

Dr. Zurqani's editorial and peer-review activities further highlight his commitment to advancing scientific research. He serves on the Editorial Boards of several journals, including *Frontiers in Earth Science* (Geoinformatics), *Discover Soil* (Springer Nature), *Frontiers in Remote Sensing*, and *Remote Sensing* (MDPI). He has also been appointed as a Guest Editor for special issues in several journals, focusing on topics such as advancing spatial prediction of soil properties using remotely sensed data and geospatial artificial intelligence, and UAV agricultural management. Dr. Zurqani's dedication to his field is further demonstrated by his role as a peer reviewer for numerous publishers, including Springer Nature, Nature, Frontiers, Elsevier, and MDPI, among others. He is also the editor of several influential publications that address critical environmental challenges. His notable works include: *The Soils of Libya* (2021), part of Springer's World Soils Book Series; *Environmental Applications of Remote Sensing and GIS in Libya* (2022); and *Water Resources of Libya: Challenges and Management* (2025), published as part of the Springer Water Series (SPWA). These publications highlight his multidisciplinary expertise and commitment to addressing top environmental issues in Libya and beyond.

Contributors

Ismail Ageena Department of Geography, Faculty of Arts, University of Al-Zawiya, Al-Zawiya, Libya

Abdulsalam Al-Bukhari Department of Forest and Rangeland, Faculty of Natural Recourses and Environmental Sciences, University of Omar Al-Mukhtar, Al Bayda, Libya

Mhi Eldin M. Alkhboli Department of Soil and Water Sciences, Faculty of Agriculture, University of Omar Al-Mukhtar, Al Bayda, Libya

Shadia A. Alzurqani College of Forestry, Agriculture, and Natural Resources, University of Arkansas at Monticello, Monticello, AR, USA

Ali A. Bataw Department of Zoology, Faculty of Sciences, University of Omar Al-Mukhtar, Al-Bayda, Libya

Murad Ellafi Department of Soil, Water, and Climate, University of Minnesota, Twin Cities, MN, USA;

Northwest Research and Outreach Center, University of Minnesota, Crookston, MN, USA

Abdulazim S. Fadil Agriculture Research Center of Libya, Al Bayda, Libya

Rod Fensham School of Biological Sciences, Faculty of Science, University of Queensland, Brisbane, Australia

Nagib Froja Soil and Water Department, Faculty of Agriculture, University of Tripoli, Tripoli, Libya

Salah Hamad Faculty of Natural Resources and Environmental Science, Omar Al Mukhtar University, Al Bayda, Libya

Salah Hamed Department of Forest and Rangeland, Faculty of Natural Recourses and Environmental Sciences, University of Omar Al-Mukhtar, Al Bayda, Libya

Mohamed A. Hamilla Irrigation and Drainage Specialist, Tripoli, Libya

Fawzi Saeid Technical Center for Groundwater Management (TCGM), Great Man-Made River Project Authority (GMMRA), Benghazi, Libya

Ali A. Shaki Department of Environmental Science, Faculty of Environment and Natural Resources, Wadi Al-Shatti University, Wadi Al-Shatti, Libya

Hamdi A. Zurqani University of Arkansas Agricultural Experiment Station, Arkansas Forest Resources Center, College of Forestry, Agriculture, and Natural Resources, University of Arkansas, Monticello, AR, USA

Introduction to the "Water Resources of Libya: Challenges and Management"

Hamdi A. Zurqani

Abstract Water is an essential and irreplaceable resource that sustains life and ecosystems on Earth. The availability and management of water resources are of primary concern in arid and semi-arid regions such as Libya because of their significant effects on various sectors, including agriculture, industry, domestic use, and the environment. This chapter introduces the book "Water Resources of Libya: Challenges and Management" and summarizes all the chapters presented in this book. The chapters of this book focus on a range of water resources management challenges in the country, such as water scarcity, quality management, infrastructure development and maintenance, governance and policy, climate change impacts and adaptation strategies, stakeholder engagement and collaboration, and sustainable use and conservation practices. These issues are thoroughly explored to provide readers with an in-depth understanding of the complexities and potential solutions to ensure appropriate management of water resources in the region. Each chapter provides insights into a different aspect of water management, covering nearly every part of preservation strategies to preserve water quality and ensure equitable distribution among users. In this book, experts and scientists share their expertise and experience in applying top-notch discoveries and suggestions to manage and improve the quality of water resources in Libya. The book aims to serve as a reliable source for academics, decision-makers, and anyone interested in using it to assess water resources and management in the country.

Keywords North Africa · The Great Man-Made river · Nubian sandstone aquifer system · Desert oases · Water management strategies · Water consumption · Climate change impacts

H. A. Zurqani (✉)
University of Arkansas Agricultural Experiment Station, Arkansas Forest Resources Center, College of Forestry, Agriculture, and Natural Resources, University of Arkansas, Monticello, AR, USA
e-mail: Zurqani@uamont.edu; Hzurqani@uark.edu

H. A. Zurqani (ed.), *Water Resources of Libya*, Springer Water,
https://doi.org/10.1007/978-3-031-80920-0_1

1 Libya: A Brief Background

Libya is a North African country bordered by the Mediterranean Sea to the north, Tunisia and Algeria to the west, Niger and Chad to the south, and Sudan and Egypt to the east. Its location is approximately between longitude 9° and 25° E and latitude 18° and 33° N (Fig. 1). With a population of over 7.3 million people, this country is the fourth largest in Africa by geographic area (The World Factbook 2024), covering approximately 1.76 million km^2 (Zurqani et al. 2019). Libya has a diverse climate, with the coastal region experiencing a Mediterranean climate characterized by mild, wet winters and hot, dry summers. With sweltering summers and moderate winters, the climate is harsher and drier in the interior regions (Zurqani 2022). With approximately 93% of the land surface receives an average annual rainfall of less than 100 mm, with most of it falling between November and March, and no perennial rivers or freshwater lakes, the country has a significant reliance on groundwater as a result of the limited access to surface water supplies (FAO 2022).

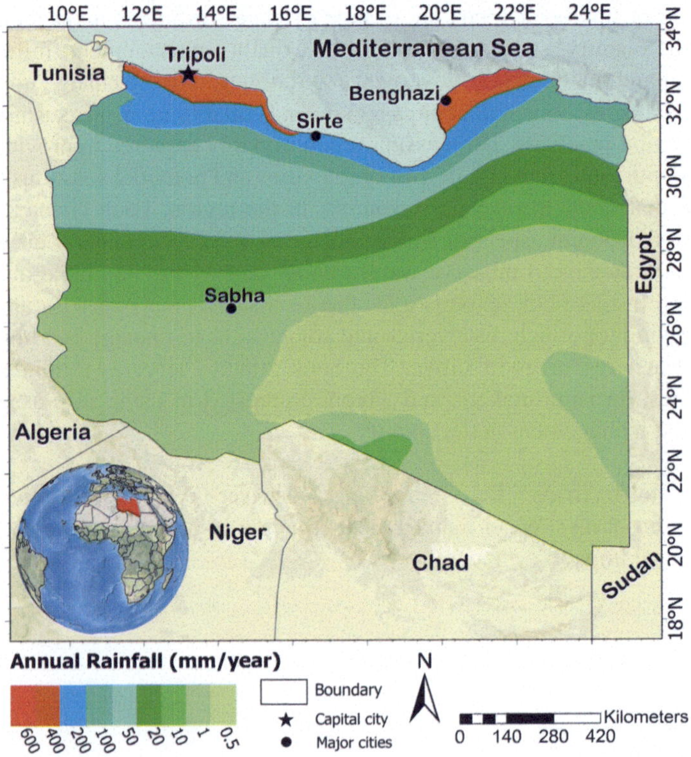

Fig. 1 Map of Libya with capital cities and the spatial distribution of annual rainfall (mm/year)

2 Water Resources: Overview

Water resources refer to sources of water that are considered valuable or potentially useful for human activities such as agriculture, industry, household, recreation, and the environment (United Nations. 2020). Arid and semi-arid regions such as Libya often face water scarcity and ecosystem degradation due to overuse and mismanagement of limited water resources (FAO 2019). Factors such as population growth and rapid urbanization contribute to a steep increase in demand for water resources, which are threatened by diminishing water quality caused by pollution, reduced quantity due to overexploitation, and increased demand for agricultural use (Population Reference Bureau 2020). Water scarcity is a primary concern worldwide, especially for the agricultural sector, which heavily depends on water supplies (Adama et al. 2020). Water scarcity occurs when demand for freshwater exceeds availability in a given area and throughout all sectors of the economy, which can threaten sustainable livelihoods that rely on water and agriculture (Adama et al. 2020; Vorosmarty et al. 2000).

Numerous studies have reported that the Sahara turned into a desert due to climate change over centuries, and many of the aquifers underneath have not been replenished with water for over 5000 years (Edmunds 2009; Engelhardt et al. 2013; Mokadem et al. 2018). However, according to data compiled from hydrogeological maps of various national governments and 152 publications that contain summaries of 283 aquifers, it has been found that Africa, and specifically Libya, has a significant amount of groundwater that could be used as a primary source of drinking water (MacDonald et al. 2012). This study also reported that the total groundwater storage in Africa is of the order of 0.66 million km^3 (0.36–1.75 million km^3), with the best estimated groundwater storage for Libya being 99,500 km^3 (ranging from 64,600 to 234,000 km^3) (Fig. 2). However, not all of this groundwater storage is available for abstraction, and the estimated volume is over 100 times the estimates of annual renewable freshwater resources in Africa.

Libya is primarily a desert country with few surface water resources (FAO 2022). The country is almost entirely dependent on groundwater, the majority of which is "fossil" water that was recharged thousands of years ago when the climate in the area was wetter (Issar 1994). Groundwater is the primary water source for domestic and non-domestic purposes in Libya, fulfilling more than 80% of the total water demand in the country. Libya has six main basins, each formed from groundwater aquifers that differ in geological, hydrogeological, and hydrochemical properties. Groundwater can also be classified into three zones: zone 1 is in the western part of Libya, zone 2 is in the eastern part, and zone 3 is in the northeastern part of the country (Fig. 3) (Mohamed 2019).

The Kufra and Sirte basins are in eastern Libya. The Kufra basin extends through southeastern Libya and northeastern Chad and comprises Cambrian to Lower Cretaceous sedimentary rocks (Wright et al. 1982). The Nubian Sandstone forms its freshwater-bearing formations, which create a semi-confined to unconfined aquifer system. The Sirte basin contains sediments ranging from the Cambrian to the Tertiary

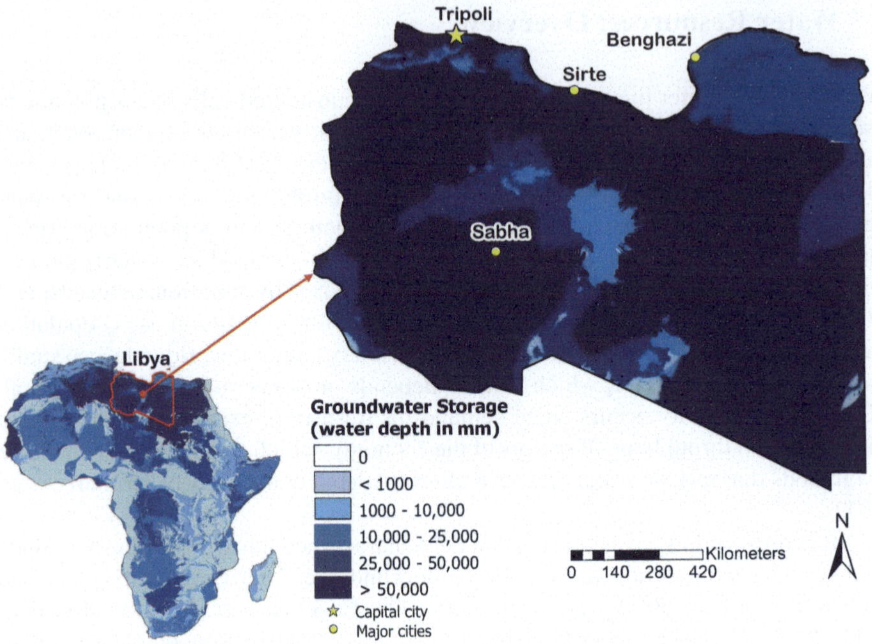

Fig. 2 Groundwater storage for Libya (adapted from: MacDonald et al. 2012)

period. The Tazerbo wellfield is located above an uplift that separates the Sirte and Kufra basins. The Sarir wellfield is represented by the Marada aquifer of the Lower to Middle Miocene age. The Murzuk basin is in southwestern Libya and comprises Cambrian to Cretaceous sedimentary rocks. The Jefara Plain is in the northern continental margin of Africa. The Hamadah (Ghadamis) basin is situated in northwestern Libya. Al-Jabal Al-Akhdar area has sedimentary rocks ranging from the Upper Cretaceous to the Tertiary periods (Mohamed 2019). These rocks were formed on the southern margin of the Tethys Sea and are composed of limestones interspersed with dolomites and marls. The sediments underwent moderate folding during the intra-Senonian and early Ypresian periods and slight warping during the Middle Eocene to Middle Miocene era (Sahli 1994).

The use of groundwater resources in Libya has been a subject of concern due to its depletion and population growth (Wheida and Verhoeven 2007). The excessive exploitation of groundwater in Libya causes a significant threat to the long-term sustainability of the country's water resources and agricultural production (Hamed et al. 2018). These unplanned human activities and increasing economic developments applied enormous stresses on groundwater resources in Libya (Abdudayem and Scott 2014). In 2008, an estimation was conducted to predict the future water consumption in Libya, which included agricultural, domestic, and industrial uses for the years 2006 to 2020 (Hamad et al. 2017). The evaluation showed that the total water consumption is expected to increase from 6293.89 million m^3 in 2006

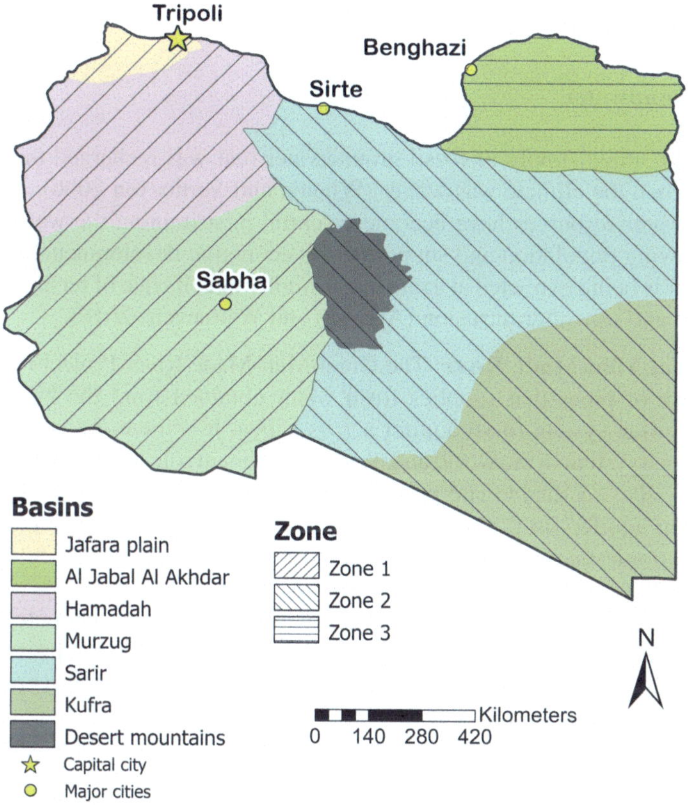

Fig. 3 Libyan basins and zones (adapted from: Mohamed 2019)

to 12,473.20 million m^3 in 2020, with a compound annual growth rate of 4.97% (Lawgali 2008). This means water consumption increased by 98% in 2020 compared to 2006. In Libya, the main sources of water are groundwater, surface water, desalinated seawater, and recycled wastewater. Groundwater accounts for almost 95% of Libya's water supply, while surface water only makes up 2%, including rainwater and water stored in dams. Desalinated seawater provides another 2%, and wastewater recycling accounts for 1% (Hamad et al. 2017). To manage the intermittent water resources in the wadis, a total of 18 dams were built to collect approximately 61 million cubic meters (MCM) of runoff water (General Water Authority 2014; Noureddine et al. 2021). In addition, there are natural springs in Jabal al-Akhdar, Jabal Nafusa, and various locations in central Libya that serve as another source of surface runoff (General Water Authority 2014; Noureddine et al. 2021). These springs have a discharge rate that ranges from 1 L per second (L/s) to over 10 L/s and are used for various purposes (General Water Authority 2014; Noureddine et al. 2021).

2.1 *The Main Water Sources in Libya*

2.1.1 Groundwater

Libya has an ample supply of fossil groundwater that was recharged thousands of years ago when the climate was wetter (Wheida and Verhoeven 2006). These non-renewable water resources have become essential in sustaining the water needs of Libya's growing population and supporting its economic development. As a result, groundwater is being extracted at a rate that is higher than its rate of recharge, leading to depletion and saltwater intrusion (Alfarrah and Walraevens 2018).

- **The Great Man-Made River**: The Great Man-Made River Project (GMMR) is a network of pipes that supplies fresh water obtained from the Nubian Sandstone Aquifer System fossil aquifer across Libya. It is the most extensive water conveyance system in the world and plays a significant role in addressing the water scarcity issues in Libya. The project aimed to deliver high-quality fresh water from deep aquifers in the south (i.e., Sahara) to the northern populated coastal cities for domestic use, agriculture, and industry (Thomas 2011). The GMMR is an alternative solution to the water shortage problem in Libya, where minimal rainfall and a rapidly increasing population have posed significant challenges to the water supply. The project was completed in 1991, and now it provides about 80% of Libya's drinking water but is also a significant contributor to groundwater depletion (Kuwairi 2006).
- **Desert Oases**: Oases in Libya are an essential water source for the desert region, providing scarce water resources for agriculture and human consumption (Zurqani 2022). The oases are scattered across the country, with the largest oases being Ghadames, Sabha, Murzuk, and Al Jawf. Notably, Al Jawf ranks as the biggest oasis in the Al-Kufra basin. These oases support a small population and some agriculture projects, but the water supply is limited and subject to seasonal variations (Fensham et al. 2023).

2.1.2 Desalination

Desalination removes salt from seawater to produce fresh water (Micale et al. 2009). It is a relatively expensive way to produce water, but it is becoming increasingly essential in Libya as groundwater resources are depleted (Ashour and Ghurbal 2004). Libya has a few desalination plants, which provide a small but growing share of the country's water supply (Elhassadi 2008).

2.1.3 Seasonal Surface Water "Rainwater Harvesting"

Rainwater harvesting is a traditional water conservation practice in Libya, where rainwater is collected and stored for various uses, such as irrigation and drinking water

(Brika 2019). However, the existing rainwater harvesting methods in Libya mainly focus on stormwater restoration and recycling without fully considering its other benefits, such as providing an alternative water source during drought periods. This practice helps to reduce dependence on non-renewable water sources and mitigate the impact of water scarcity in the region (Hejazi et al. 2023).

Libya faces a severe water crisis due to limited resources and ongoing economic and political instability (Zurqani 2022). These factors have put significant pressure on groundwater resources and led to shortages in the water supply. Thus, sustaining water resources in Libya has become a deep concern for decision-makers and scientists due to the increasing economic developments and unplanned human activities that have put immense stress on groundwater resources. International aid and cooperation are essential to address the water crisis in the country and ensure sustainable access to clean water for the population. The Libyan government has adopted a long-term planning view that can potentially address the severe water crisis in the country. These efforts include:

- Reducing water consumption in the agricultural sector
- Increasing the use of water-efficient irrigation techniques
- Expanding the use of desalination
- Promoting rainwater harvesting.

Despite these efforts, water scarcity will likely continue to be challenging in Libya due to limited resources and increasing demand. As a result, the country must prioritize investments in water conservation and management measures to ensure a sustainable water supply for its population.

3 Challenges in Water Resources Management

The water resources in Libya are heavily dependent on non-renewable sources, leading to concerns about long-term sustainability. In addition, the political instability in the region and lack of investment in water infrastructure further compound the ability to manage water resources effectively. Despite the presence of the Great Man-Made River and underground aquifers, managing water resources in Libya has serious challenges for sustainable development. These challenges include:

3.1 Depletion of Aquifers

The intensive use of groundwater for agricultural, industrial, and domestic use has led to the depletion of aquifers at an unsustainable rate especially in the northern coastal aquifers. This raises concerns about the long-term availability of water and the risk of land subsidence from underground cavity collapse.

3.2 *Population Growth*

Rapid population growth, urbanization, and increased industrialization have unprecedentedly increased water consumption, creating a critical demand for freshwater. Thus, as the population continues to grow, it is expected to put increasing pressure on existing water resources in the coming years.

3.3 *Climate Change*

The impacts of climate change, which include changes in precipitation patterns and increased temperatures, exacerbate Libya's water problems. Water availability can be disrupted, and increased evaporation rates and unusual periods of low precipitation aggravate drought conditions.

3.4 *Conflict and Political Instability*

Political instability and conflict have also affected the management of water resources in Libya. The Great Man-Made River and other water infrastructure are unable to function properly due to maintenance and infrastructure problems, leading to severe limitations in water availability for both domestic and agricultural purposes.

4 Water Management Strategies

In Libya, efficient and sustainable management of water resources is essential for future development and stability. Without proper management and conservation efforts for the limited water reserves, the country will continue to be at risk of severe water shortages and droughts. Furthermore, the strong population growth and economic developments in Libya aggravate the groundwater resources stress due to the expanding agricultural, industrial, and municipal needs. Accordingly, the amount of water being used far exceeds the water supply, leading to severe shortages and depletion of water resources. Several water management strategies can be used to address the challenges mentioned above:

4.1 Water Conservation and Efficiency

Implementing water-saving technologies in agriculture, improving irrigation practices, and promoting water-efficient technologies in industries and households can significantly reduce water consumption.

4.2 Alternative Water Sources

Exploring alternative water sources, such as seawater desalination, can provide a supplementary source of freshwater, especially for coastal areas. However, desalination comes with its own set of environmental and economic challenges.

4.3 Integrated Water Resource Management

Adopting an integrated approach that considers the interconnectedness of surface water and groundwater systems can lead to more sustainable water management. This includes regulating groundwater extraction, managing surface water runoff, and protecting water quality.

4.4 Renewable Energy Integration

The energy-water nexus is critical in water-stressed regions. Integrating renewable energy sources into water infrastructure, such as using solar power for desalination or pumping, can reduce operational costs and environmental impact.

4.5 Policy and Governance

Developing clear and effective water policies and robust governance structures can help manage water resources more sustainably. International collaboration and knowledge sharing can also provide valuable insights into effective water management practices.

5 Summaries of the Book's Chapters

This chapter aims to introduce the book by summarizing each chapter. Chapter 2 is titled "**The Great Man-Made River Project in Libya: An Overview of the Current Status and Future Prospects**". It gives an overview of the Great Man-Made River Project (GMMRP), including its current status and future prospects. The GMMRP was implemented to address the water scarcity problem in the country by supplying fresh groundwater to northern Libyan cities for residential, agricultural, and industrial uses. It successfully addressed the country's water shortage using a cost-effective and sustainable method. Thus, the project achieved its goals but faced challenges due to the civil war, including financial difficulties, infrastructure problems, and unclear water allocation. The text recommends solutions to ensure the project's sustainability, such as improved financial management, better infrastructure maintenance, and a national water resources strategy.

Chapter 3 focuses on "**Groundwater: An Important Resource of Urban Water Supply in Libya**". Libya faces a severe scarcity of renewable water resources, with each person having access to less than 10% of the global average. Groundwater is the primary water source in Libya, accounting for 97% of the country's total water consumption. The Great Man-Made River Project (GMMRP) is currently the primary source of water supply for more than 70% of Libyan cities. Still, it cannot meet the projected water demand, which is expected to be around 10 billion m3 by 2035. Other sources, such as sewage treatment and saltwater desalination, must be developed to close the gap between supply and demand. Water supply services in urban areas face severe challenges due to aging facilities, ongoing armed conflict, political, economic, and institutional instability, and continuous power and fuel supply cuts.

In Chap. 4, the authors discuss the **"Rainwater Harvesting in Libya Sustainable Solutions for Water Scarcity"**. Rainwater harvesting is a viable solution to address water scarcity challenges in arid regions such as Libya. This chapter explores the potential of rainwater harvesting in Libya, where water resources are limited due to low rainfall and reliance on finite sources. Traditional methods like rooftop collection and modern techniques such as dam construction offer opportunities to augment water resources sustainably. However, ensuring the safety and reliability of harvested rainwater is essential, necessitating comprehensive water quality testing and treatment protocols. Despite challenges like low rainfall and high evaporation rates, innovative solutions based on modern technologies and sustainable practices can enhance water resilience. Selected case studies in this chapter illustrate practical implementation and outcomes, providing valuable insights for plans. Recommendations include allocating resources for infrastructure development, educating communities, establishing water quality protocols, prioritizing dam construction, strengthening capacity, and formulating supportive policies. By implementing these recommendations, Libya can effectively address water scarcity challenges and ensure water security for present and future generations.

On the other hand, Chap. 5 discusses **"Agricultural Drainage in Libya: History and Perspective"**. This chapter discusses agricultural drainage in Libya from the

1950s to the present day. It reviews the development of drainage systems, analyzes the existing systems, and proposes solutions for improvement. In the 1950s, there were no drainage systems in place, and farms suffered from waterlogging and salinity. By the 1960s, some agricultural projects were underway, but drainage was often neglected or poorly maintained, resulting in a decrease in land productivity. In 1985, the General Administration of Irrigation and Drainage (GAID) was established. The GAID has focused on improving irrigation and drainage systems throughout Libya. They have identified several problems in Libya's agricultural drainage systems, including inappropriate designs, salt accumulation in the soil, poor irrigation water quality, and a shallow water table. The chapter also details specific agricultural projects in Libya, including Ain Tawergha, Al-Hammam, Al-Furjan, Zillah, and Brak Eshkeda. These projects have all faced challenges with drainage and salinity, and some have had drainage systems installed. The chapter concludes by proposing solutions to improve drainage in these areas.

Furthermore, Chap. 6 of the book provides an overview on the topic of **"Irrigation Development in Libya: Review and Assessment"**. This chapter reviews and assesses irrigation development in Libya from the 1960s to the 2010s. It highlights the challenges the country faces in agricultural irrigation and proposes solutions for improving it. Water scarcity is a significant problem in Libya due to a shortage of water resources and increased demand for industrial, domestic, and agricultural expansion. The lack of maintenance programs for irrigation systems has resulted in low efficiency and early corrosion of tools. This issue has hindered the expansion of agriculture, leading to productivity losses and impacting development in irrigated lands. The chapter discusses various irrigation systems used in Libya, including sprinklers, central pivots, gun sprinklers, one-sided move sprinklers, hanged sprinklers, and localized irrigation systems. Surface irrigation systems, including basin, furrow, and border irrigation, are also addressed. The authors propose several solutions for improving irrigation water management and raising standards in this sector. These include providing financial and technical resources, developing human capacities in irrigation, and making technical information and specialized scientific references more accessible. They conclude that improving irrigation systems in Libya is necessary for agriculture and food security.

In addition, Chap. 7 **"Irrigation Water Resources in Libya: Challenges and Potentials"** explores the challenges and opportunities related to irrigation in Libya. Libya faces challenges securing water resources for irrigation due to its arid climate and limited freshwater reserves. Most irrigation water in Libya is derived from groundwater sources, which are being depleted at an alarming rate. This overuse of groundwater for agriculture has led to concerns about salinity and seawater intrusion into coastal aquifers. This chapter highlights the importance of sustainable water management practices, such as drip irrigation and rainwater harvesting, to conserve water resources. It also emphasizes the need for improved monitoring and regulation of groundwater use to ensure long-term sustainability. The chapter concludes by calling for a multi-pronged approach to address the challenges of water scarcity in Libya. This approach includes investing in water conservation measures, developing alternative water sources, and promoting public awareness about the importance of water conservation.

In Chap. 8 with the title **"An Overview of Flood Hazards in Libya: Impacts and Required Actions"** the authors discuss the causes, impacts, and required actions to reduce flood risk in the country. Floods are a natural hazard that can devastate human health, infrastructure, and the environment. Heavy rainfall, dam failures, or climate change can cause them. Floods are most common in coastal areas, dry riverbeds (wadis), and the highland regions. They can cause widespread damage and displacement and impact cultural heritage, human mobility, and the lives and livelihoods of local communities. The lack of available data makes it challenging to estimate floods' frequency and impact on the country accurately. The chapter concludes by emphasizing the need for a comprehensive approach to flood risk management in Libya. This includes measures such as improved early warning systems, investment in infrastructure, and promoting sustainable land use practices.

Chapter 9, which is entitled **"Future of Water Supply and Demand in Libya under Climate and Land-Use Change"** examines the relationship between climate patterns and water supply and demand in Libya over 60 years, from 1961 to 2020. This chapter analyzes climate change trends in Libya from 1961 to 2020, focusing on temperature, precipitation, and evapotranspiration. The findings of this chapter reveal a significant warming trend, particularly between July and October, with notable temperature increases in several locations. Precipitation in Libya is erratic, and evapotranspiration exhibits regional variations. Water availability is a critical concern due to its arid climate and limited freshwater resources, and agriculture is the leading consumer of water. Sustainable land management practices are important in addressing climate change, with deforestation and urbanization contributing to around 23% of global greenhouse gas emissions. This chapter also provides valuable insights into climate change trends in Libya. It emphasizes the importance of understanding these trends for informed decision-making and strategies to mitigate the environmental and socio-economic impacts of climate change.

The book ends with the conclusions and recommendations Chap. 10 which is entitled **"Water Resources of Libya: Challenges and Management"**. This chapter outlines recommendations for addressing the water resource challenges in Libya. Libya faces significant challenges in managing its water resources due to population growth, climate change, and inadequate infrastructure. These factors have contributed to limited water resources availability in the country, which has led to a heavy reliance on groundwater as a primary source of drinking water. However, groundwater extraction for agricultural and domestic purposes causes several challenges, including over-extraction, contamination, and declining water tables. This chapter highlights the most important conclusions and recommendations from all chapters to provide perspective and guidance for students, teachers, researchers, decision-makers, etc. It includes summarized information to provide readers with an in-depth understanding of the complexities and potential solutions to ensure appropriate management of water resources in the region.

6 Conclusion

Libya experiences critical water stress, and according to a 2019 study by the Water Resources Institute (WRI), a global research organization based in Washington, D.C., United States, Libya was ranked as the sixth most water-stressed country globally (UNICEF 2022; WRI 2024). The country has a relatively high-water consumption rate, and its population is growing rapidly. Most of the water extractions are unsustainable, as they are from groundwater sources. This has increased the demand for water resources in the country and created several other challenges. Overuse of groundwater, which is causing saltwater intrusion and a decline in water levels, is one of the main problems in the country. Another challenge Libya faces is water pollution caused by various sources, such as sewage, agricultural runoff, and industrial waste. This chapter provides an overview of the book "Water Resources of Libya: Challenges and Management", which discusses many challenges in managing the water resources of the country. By combining conservation efforts, cutting-edge technologies, and integrated management techniques, Libya has the potential to overcome these problems and pave the way for water security for both its present and future generations. However, cooperation and coordination among governmental and local organizations and international partners will be necessary to implement a sustainable water management system. This coordinated approach is essential for addressing the country's complex water management issues, such as water demand, quality assurance, and water resource preservation.

References

Abdudayem A, Scott AH (2014) Water infrastructure in Libya and the water situation in agriculture in the Jefara region of Libya. Afr J Econ Sustain Dev 3(1):33–64

Adama GJ, Jimoh DO, Otache MY (2020) Optimization of irrigation water allocation framework based on genetic algorithm approach

Alfarrah N, Walraevens K (2018) Groundwater overexploitation and seawater intrusion in coastal areas of arid and semi-arid regions. Water 10(2):143

Ashour MM, Ghurbal SM (2004) Economics of seawater desalination in Libya. Desalination 165:215–218

Brika B (2019) The water crisis in Libya: causes, consequences and potential solutions. Desal Water Treat. https://www.deswater.com/DWT_articles/vol_167_papers/167_2019_351.pdf

Edmunds WM (2009) Palaeoclimate and groundwater evolution in Africa—implications for adaptation and management. Hydrol Sci J 54(4):781–792

Elhassadi A (2008) Horizons and future of water desalination in Libya. Desalination 220(1–3):115–122

Engelhardt I, Rausch R, Keim B, Al-Saud M, Schüth C (2013) Surface and subsurface conceptual model of an arid environment with respect to mid-and late Holocene climate changes. Environ Earth Sci 69:537–555

Fensham RJ, Adinehvand R, Babidge S, Cantonati M, Currell M, Daniele L, Elci A, Galassi DM, de la Hera Portillo Á, Hamad S, Han D (2023) Fellowship of the Spring: an initiative to document and protect the world's oases. Sci Total Environ 887:163936

Food and Agriculture Organization of the United Nations (FAO) (2019) The state of the world's land and water resources for food and agriculture. Rome. https://www.fao.org/3/i1688e/i1688e.pdf
Food and Agriculture Organization of the United Nations (FAO) (2022) Libya water scarcity and climate change. Food and Agriculture Organization of the United Nations, Rome. https://www.unicef.org/mena/media/19321/file/Libya%20water%20scarcity%20analysis%20and%20recommendations_%20UNICEF%20Sep%202022.pdf
General Water Authority (2014) Water and energy for life in Libya (WELL), Project funded by the European Commission No. 295143, FP7, Libya
Hamad JRJ, Hanafiah MM, Yaakob WZW (2017) Water resources management in Libya: challenges and future prospects. Malay J Sustain Agric 1(2):2–5
Hamed Y, Hadji R, Redhaounia B, Zighmi K, Bâali F, El Gayar A (2018) Climate impact on surface and groundwater in North Africa: a global synthesis of findings and recommendations. Euro-Mediterr J Environ Integr 3:1–15
Hejazi M, Santos Da Silva SR, Miralles-Wilhelm F, Kim S, Kyle P, Liu Y, Vernon C, Delgado A, Edmonds J, Clarke L (2023) Impacts of water scarcity on agricultural production and electricity generation in the Middle East and North Africa. Front Environ Sci 11:157
Issar AS (1994) Fossil water under the deserts of the Middle East: a safeguard against strife. Stud Environ Sci 58:351–361
Kuwairi A (2006) Water mining: the Great Man-made River, Libya. Proc Inst Civ Eng—Civ Eng 159(5):39–43. https://doi.org/10.1680/cien.2006.159.5.39
Lawgali FF (2008) Forecasting water demand for agricultural, industrial and domestic use in Libya. Int Rev Bus Res Papers 4(5):231–248
MacDonald AM, Bonsor HC, Dochartaigh BÉÓ, Taylor RG (2012) Quantitative maps of groundwater resources in Africa. Environ Res Lett 7(2):024009
Micale G, Cipollina A, Rizzuti L (2009) Seawater desalination for freshwater production. In: Seawater desalination: conventional and renewable energy processes, pp 1–15
Mohamed A (2019) Hydro-geophysical study of the groundwater storage variations over the Libyan area and its connection to the Dakhla basin in Egypt. J Afr Earth Sc 157:103508
Mokadem N, Redhaounia B, Besser H, Ayadi Y, Khelifi F, Hamad A, Hamed Y, Bouri S (2018) Impact of climate change on groundwater and the extinction of ancient "Foggara" and springs systems in arid lands in North Africa: a case study in Gafsa basin (Central of Tunisia). Euro-Mediterr J Environ Integr 3:1–14
Noureddine G, Eslamian S, Katlane R (2021) Integrated water resources management and climate change. Int J Water Sci Environ Technol VI(1):89–103
Population Reference Bureau (2020) Finding the balance: population and water scarcity in the Middle East and North Africa. https://www.prb.org/resources/finding-the-balance-population-and-water-scarcity-in-the-middle-east-and-north-africa/
Sahli NI (1994) Water Quality of the Sarir Wellfield. University of Technology, Sydney, Libya. Project Rept for GRAD Dipl. In Engineering
The World Factbook (2024) Libya. https://www.cia.gov/the-world-factbook/countries/libya/
Thomas F (2011) The GMMR Project: Libya's achievement and NATO as war criminal party-pooper, OpEdNews, 01 Sep 2011 https://www.derechos.org/peace/libya/doc/lby1035.html
United Nations International Children's Emergency Fund (UNICEF) (2022) Libya water scarcity and climate change: © UNICEF Libya an analysis on WASH enabling environment in Libya. https://www.unicef.org/mena/documents/unicef-libya-water-scarcity-and-climate-change-analysis-wash-enabling-environment-libya#:~:text=Libya%20experiences%20critical%20water%20stress,most%20water%20stressed%20country%20globally
United Nations (2020) The state of the world's water resources. United Nations, New York. https://www.unwater.org/publications/un-world-water-development-report-2020
Vorosmarty CJ, Green P, Salisbury J, Lammers RB (2000) Global water resources: vulnerability from climate change and population growth. Science 289(5477):284–288

Water Resources Institute (WRI) (2024) 25 Countries, housing one-quarter of the population, face extremely high water stress. https://www.wri.org/insights/17-countries-home-one-quarter-worlds-population-face-extremely-high-water-stress
Wheida E, Verhoeven R (2006) Review and assessment of water resources in Libya. Water Int 31(3):295–309
Wheida E, Verhoeven R (2007) An alternative solution of the water shortage problem in Libya. Water Resour Manage 21:961–982
Wright EP, Benfield AC, Edmunds WM, Kitching R (1982) Hydrogeology of the Kufra and Sirte basins, eastern Libya. Q J Eng GeolHydrogeol 15(2):83–103
Zurqani HA (2022) Introduction to environmental applications of remote sensing and GIS in Libya. In: Environmental applications of remote sensing and GIS in Libya, pp 1–9. Springer International Publishing, Cham
Zurqani HA, Mikhailova EA, Post CJ, Schlautman MA, Elhawej AR (2019) A review of Libyan soil databases for use within an ecosystem services framework. Land 8(5):82

Dr. Hamdi A. Zurqani is an Assistant Professor of Geospatial Science in Natural Resource Management and Conservation at the University of Arkansas Agricultural Experiment Station, Arkansas Forest Resources Center, University of Arkansas at Monticello, USA. With a strong academic foundation and a passion for innovative technologies, he is also an FAA (i.e., Pt107) Certified sUAS/Drones Pilot, leveraging this expertise to enhance his proficiency in remote sensing, GIS, and geospatial data analysis, which he applies to advance his research and teaching in environmental monitoring and analysis. Dr. Zurqani is a recognized expert as a result of his internationally acclaimed work in the areas of environmental information science, remote sensing, geospatial analysis, land evaluation, sustainability, pedology, and soil science education. He has conducted research across the world, including the United States of America and Africa, and has served as PI, co-PI, or co-investigator on several grants-funded research projects. In addition to his research and teaching responsibilities, Dr. Zurqani is actively involved in institutional service at the University of Arkansas at Monticello. He serves on the UAM Campus-wide Artificial Intelligence (AI) Review Committee, the UAM Faculty Assembly Student Engagement Committee, and the UAM Library Committee.

Professionally, Dr. Zurqani contributes to the broader academic community through a variety of leadership and collaborative roles. He is an active member of the Steering Committee of the Institute of Electrical and Electronics Engineers (IEEE) Mediterranean and Middle-East Geoscience and Remote Sensing Symposium (M2GARSS) and serves as an Ambassador for the Geoscience and Remote Sensing Society (GRSS) Chapter of the IEEE in Libya, facilitating communication, collaboration, and knowledge-sharing within the geospatial and remote sensing communities. In recognition of his expertise and substantial contributions to the field, Dr. Zurqani was invited in 2024 to serve on the Technical Program Committee for the IEEE Asia-Pacific Conference on Geoscience, Electronics, and Remote Sensing Technology (AGERS). In 2025, he also joined the IEEE GRSS Technical Committee on Remote Sensing Environment, Analysis, and Climate Technologies (REACT), a platform for scientists and engineers to exchange ideas and knowledge on environmental issues and the impacts of climate change, emphasizing support for sustainable development goals (SDGs) through remote sensing technologies. Beyond these roles, Dr. Zurqani holds the position of UAM Co-Principal Investigator at the Southeast Climate Adaptation Science Center, where he is involved in research initiatives focused on climate adaptation and resilience in the southeastern United States and its surrounding regions.

Dr. Zurqani's editorial and peer-review activities further highlight his commitment to advancing scientific research. He serves on the Editorial Boards of several journals, including *Frontiers in Earth Science* (Geoinformatics), *Discover Soil* (Springer Nature), *Frontiers in Remote Sensing*, and *Remote Sensing* (MDPI). He has also been appointed as a Guest Editor for special issues in several journals, focusing on topics such as advancing spatial prediction of soil properties using remotely sensed data and geospatial artificial intelligence, and UAV agricultural management. Dr. Zurqani's dedication to his field is further demonstrated by his role as a peer reviewer for numerous publishers, including Springer Nature, Nature, Frontiers, Elsevier, and MDPI, among others. He is also the editor of several influential publications that address critical environmental challenges. His notable works include: *The Soils of Libya* (2021), part of Springer's World Soils Book Series; *Environmental Applications of Remote Sensing and GIS in Libya* (2022); and *Water Resources of Libya: Challenges and Management* (2025), published as part of the Springer Water Series (SPWA). These publications highlight his multidisciplinary expertise and commitment to addressing top environmental issues in Libya and beyond.

The Great Man-Made River Project in Libya: An Overview of the Current Status and Future Prospects

Salah Hamad and Fawzi Saeid

Abstract The Great Man-Made River Project (GMMRP) was implemented to address Libya's water scarcity problem by utilizing groundwater instead of desalination due to cost, capacity expansion, and operational considerations. The GMMRP aimed to produce and transport 6.4 million cubic meters of groundwater daily to meet the water demands of northern Libyan urban settlements for domestic, agricultural, and industrial use. The GMMRP has achieved its objectives of improving the quality of life for residents, alleviating pressure on coastal aquifers, and promoting sustainable development, without any adverse environmental impacts. However, the GMMRP has faced significant challenges following the 2011 Libyan civil war. These include a decline in financial resources due to political conflicts, lower oil production and prices, and direct and indirect economic losses from damage caused by armed conflict and lack of security. Additionally, the GMMRP faces a considerable discrepancy between the cost of water production and the price charged to consumers, insufficient revenue generation, and an inability to import critical spare parts and materials for maintenance operations. Other challenges include unclear criteria for water allocation, incomplete information to estimate the costs of development investment plans, frequent electricity supply interruptions, and infrastructure issues. Furthermore, to ensure the project's sustainability, its recommended to develop a sustainable financial plan, improve billing and collection processes, priorities infrastructure maintenance, and develop effective monitoring and inspection systems. Also, it suggested creating a comprehensive national strategy for managing water resources and prioritizing water supplies for urban, industrial, and agricultural uses. The integration of planning for irrigation water management, investing in treated wastewater, and transparently allocating water among users based on economic, technical, social, and environmental standards are also recommended. Additionally, strengthening security measures, implementing effective pollution control measures, and promoting

S. Hamad (✉)
Faculty of Natural Resources and Environmental Science, Omar Al Mukhtar University, Al Baydah, Libya
e-mail: salah.hamad@omu.edu.ly

F. Saeid
Technical Center for Groundwater Management (TCGM), Great Man-Made River Project Authority (GMMRA), Benghazi, Libya

H. A. Zurqani (ed.), *Water Resources of Libya*, Springer Water,
https://doi.org/10.1007/978-3-031-80920-0_2

water-saving technologies and practices. Overall, the GMMRP has been successful in providing a solution to Libya's water scarcity problem. Still, there is a need for concerted efforts to address the challenges it faces to ensure its sustainability.

Keywords GMMPR · GMMRA · Libya · Groundwater · Transboundary aquifer

1 Introduction

Libya's climate is primarily arid and desert, with hot and dry summers and mild winters. The country is in North Africa and has a long coastline along the Mediterranean Sea. The northern coastal areas have a Mediterranean climate characterized by dry summers and mild winters, with most precipitation falling in the winter. In contrast, the country's interior has a tropical desert climate, with very little rainfall and high temperatures throughout the year. Al Jabal Akhdar in the east and Jabal Nafusa in the west experience cooler temperatures, and both receive the most rain in Libya. Al Jabal Al khadar in the northeast receives an annual rainfall of 400–600 mm, Jabal Nafusa in the northwest receives 200 to 400 mm. The rest of the northern parts receive 100–150 mm While the interior areas experience hot, scorching temperatures with extreme variation within a day, and rain is rare and irregular (NPC 2014; Bakhbakhi and Salem 1999; Salem 1992; Pallas 1980).

Moreover, the Sahara Desert, which covers most of Libya's interior, significantly determines the country's weather patterns. The desert's extreme temperatures and lack of vegetation contribute to hot and dry winds that can cause sandstorms and dust storms across the country. Moreover, Libya's location in the Mediterranean basin exposes it to sudden and extreme weather events, such as heavy rainfall and flooding. These events can have severe consequences, especially in urban areas where infrastructure is often inadequate to handle the sudden influx of water. However, less than 2% of the country receives enough rain (Fig. 1) to support agriculture, and only 5% receives more than 100 mm of rain annually (NPC 2014; GWA 2006; Hamad 2012a).

Libya's hydrogeology can be divided into two main regions: the northern and southern regions. The northern regions, which include the coastal plains and some inland areas, receive seasonal surface runoff and have some renewable groundwater resources. This region also has the country's major cities, including the capital, Tripoli and Benghazi, and the agricultural areas in the Jefara Plain and other coastal plains (UNECA 2005; Bakhbakhi and Salem 1999; Pallas 1980). Groundwater in the northern regions exists in several aquifers, including the Quaternary, Tertiary, Upper Cretaceous, and Triassic aquifers. These aquifers are generally recharged during seasonal surface runoff and extend over most Libyan lands bordering the Mediterranean Sea. The Quaternary aquifer underlies much of the country's coastal plains. It is composed of unconsolidated sand and gravel and is recharged during rainy seasons. This aquifer is relatively shallow and can produce large quantities of water, but its water quality is highly variable, with some areas having very saline

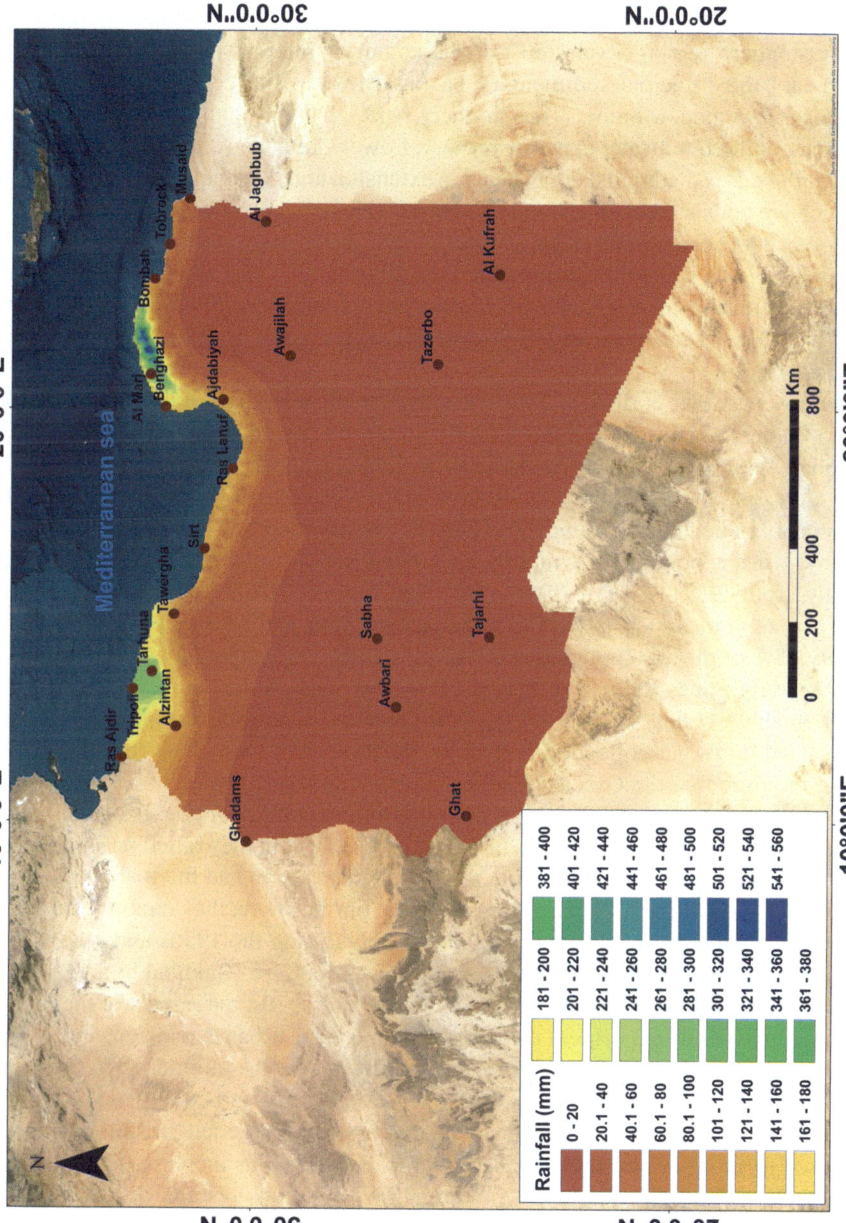

Fig. 1 Annual rainfall map of Libya (Hamad 2012a)

water. The Tertiary and Upper Cretaceous aquifers are deeper and have better water quality than the Quaternary aquifer. They are composed of consolidated rocks and are recharged by the percolation of surface water runoff (Wheida 2012; Hamad 2012a, b; Salem 1992; Shaki 2006).

The southern region covers most of the country's interior, including the Sahara Desert and the principal sedimentary basins. Most of the country's oil and gas reserves are located. Moreover, the southern region has no significant surface water resources, and groundwater exists in Tertiary, Lower Cretaceous, Triassic, and Paleozoic aquifers. It is distinguished by its wide extension and thick aquifers, the majority of which are consisted of sandstone formations (Hamad and Ahweej 2020; Binsariti and Saeed 2000; Salem 1992; Pallas 1980). The aquifers represent a massive reserve of non-renewable water that accumulated ten thousand of years ago during the Pleistocene. According to research using radioactive isotopes (carbon 14), this water dates back late Pleistocene and Holocene (Guendouz and Moulla 2010; Swailem et al. 1984).

Moreover, Libya is divided into groundwater management divisions (Groundwater systems) identified by Pallas (1980), Jones (1971); Jifarah plain, Gadamis-Al Hamadah, Murzuq, AI Jabal Al Akhdar, Sirt-Sarir and AI Kufrah. These groundwater divisions include areas that are not strictly geological basins but have been erroneously indicated as groundwater basins in some sources.

Libya's urban population is rapidly increasing, and the urbanization rate has risen steadily, from 49% in 1970 to 79% in 2020 (BCS 2006, 2020). Today, coastal urban areas in major cities such as Tripoli, Benghazi, Misrata, and Al Baydah have approximately 85% of the total population. Over time, Libya's population has experienced water scarcity issues (NPC 2014). In the past, the north of the country's population relied on shallow dug wells from shallow Quaternary aquifers, as well as natural springs and rainwater collection reservoirs, the majority of which were constructed during the time that the Greeks, Romans, and Phoenicians settled in Libya. At the same time, the southern regions relied on the water supply from oases and shallow groundwater wells (Hamad and El Hasia 2022). In addition, they relied on some relatively deep water wells drilled during the Italian occupation and the British military administration in Libya (Van Everdingen 1962). Furthermore, they continued to rely on the same water sources for urban water supply during the 1950s and 1960s and witnessed a slight increase in water supply for agriculture (Bakhbakhi and Salem 1999; Pim and Binsariti 1994). During the 1960s and 1970s, the impact of the boom in oil production and export was the population increase, urban expansion, and intensified agricultural activities; therefore, the rise in water consumption led to pressure on the northern coastal aquifers causing water level decline and quality deterioration due to seawater intrusions. However, as oil exploration proceeded, drilling activities increased to supply water to oil industry companies and thus increase the country's hydrogeology knowledge. The oil investigation operations resulted in exploration massive groundwater in southern Libya (Pim and Binsariti 1994; Pallas 1980).

The Libyan authorities were aware of the impending water crisis in Libya from an early stage. There has been a serious effort to develop additional sources of water, where the Libyan authorities decided to implement its national solution because

Libya had two options; desalination from the Mediterranean Sea or groundwater from the southern geological basins; after conducting studies on cost, system capacity expansion, and system operation, the Libyan authorities opted to utilize groundwater instead of desalination (Gijsbers and Loucks 1999). In the early 1980s plan to establish large-scale agricultural projects in southern Libya was modified to a new plan to transport groundwater from the south to densely populated coastal areas, where promising agricultural development could occur. In 1981, the Secretariat of Land Reclamation and Development began planning and conducting preliminary studies for a project initially known as the "Coastal Belt Water Supply Project." However, the project was later renamed the "Great Man-Made River Project" (GMMRP), which became the most extensive and expensive water conveyance project ever planned, designed, and constructed (see Fig. 2) (Siala and Stoner 2006; Mcloughlin 1991; Salem 1992).

The Libyan Basic Peoples Congresses (former Libyan local government) decided to implement and fund the GMMRP at the grassroots level. The General people's committee (Former Libyan Government) issued Laws No. 10/1983 and No. 11 1983 that authorizes to implementation of the GMMRP and is designed for the abstraction and transport of 6.4 million cubic meter (MCM) per day of groundwater through the project systems to cover the water demands in the northern urban settlements in terms of domestic, agricultural and industrial, in addition to the following objectives (NPC 2014; GPC 1999):

- To alleviate the pressure on the coastal aquifers in northern Libya, which have been subjected to seawater intrusion and depletion caused by the overexploitation of groundwater, which will halt further deterioration of these aquifers and enable their recovery.
- To improve the population's quality of life in the northern urban settlements by providing reliable and sufficient urban water supplies.
- To foster sustainable development by efficiently harnessing available groundwater resources, boosting agricultural productivity in the region by providing crop irrigation water. Additionally, to curb the migration of the southern population by establishing agricultural centres and supporting the growth of industries in the area by providing the necessary water resources.

This chapter aims to assess the GMMRP's current situation and prospects. The chapter will provide a historical overview of the GMMRP's inception, development, and necessity as a strategic solution to Libya's water crisis. It will also present its components, including the conveyance and distribution system, wellfields, and other GMMRP components. It also discusses the GMMRP's current situation and the most pressing challenges.

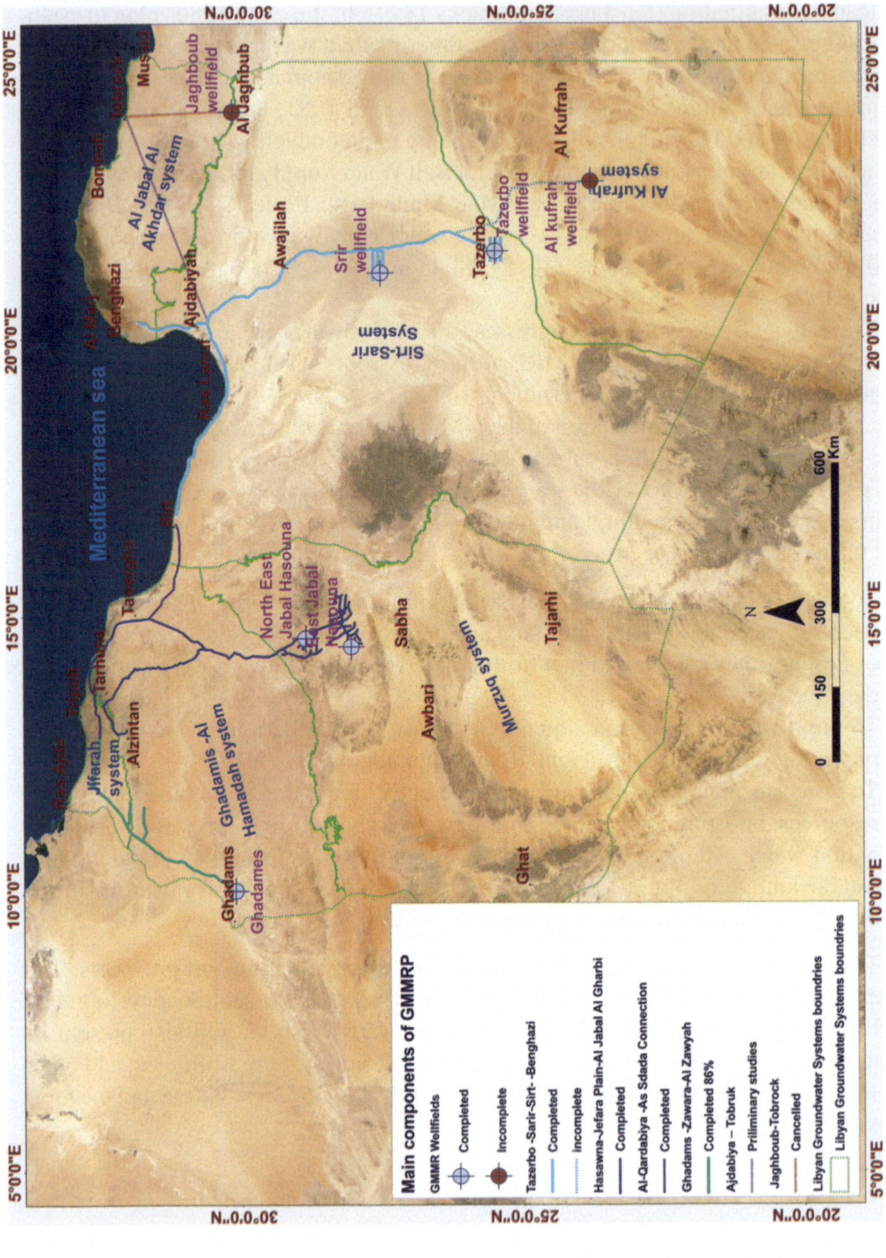

Fig. 2 Map of the main components of GMMRP map based on data from GMMRA (2022)

2 Components of GMMRP

2.1 GMMRP Eastern System

Laying the foundation stone for the GMMRP eastern system was on 28-Aug-1984 and consisted of four subsystems (Fig. 3; Table 1). The eastern system began operating partially in September 1993 with a capacity of 170,000 m^3 per day to supply cities, regions and agricultural projects with their water demands until the average daily production reached about 0.85 MCM/day. The total produced water from 1993 to 2022 was about 6079 MCM. The following subsection describes the GMMRP eastern system components.

2.1.1 Eastern System Wellfields

Sarir wellfield (Fig. 3; Table 2) consists of 159 production water wells drilled to approximately (450–478) meters. The water wells are arranged in three parallel rows (10 km apart), with 42 wells 1.3 km apart in each row and 11 deep and shallow piezometric wells monitoring the surface and deep aquifers. The wellfield was designed to produce 1 MCM/day, and groundwater abstraction commenced in 1993. Two principal aquifers are hydraulically separated; the shallow aquifer is a water table aquifer related to the post-Middle Miocene sediments, and their thickness ranges from 100 to 210 m (Calancio formation). The deep aquifer is multi-layered, developed during the Post-Middle and Lower-Middle Miocene, and behaves as a leaky confined aquifer its maximum saturated thickness penetrated by the production wells varies from 100 to 250 m (Marada formation). The depth to groundwater level ranges between 53 to 88 m. The maximum drawdown in the piezometric since the start of abstraction is 12 m. The productivity of the Sarir field wells was reduced from 101 l/s to 75 l/s due to problems with the pumps and design issues with the water well screen. To address these issues and preserve the integrity of the wells, modifications were made to the pumps. GRP well screens were installed in some wells which produce sand more than the specified rate due to manufacturing problems related to welding screen wires. To compensate for the reduced yield, 33 new water wells were drilled along the existing production water wells west of the field.

Tazerbo water wellfield is about 160 km from the southern Sarir wellfield (Fig. 4; Table 4). The wellfield developed between 1995 and 2003 consists of 108 production wells, ranging from 450 to 600 m in depth, spaced 1.3 km apart and distributed in three rows 10 km apart. Each row has two lines containing 36 water wells in addition to 35 monitoring wells scattered throughout the wellfield monitoring the surface, intermediate and deep aquifers. Deep below the drilled production water wells in the Tazerbo wellfield are three aquifers; surface, intermediate, and deep, separated by isolating layers (Aquitards). These aquifers consist mainly of sandstone. There is another even deeper aquifer (Cambro-Ordovician) not penetrated by the production water wells but intercepted by the piezometer no. 1 about 90 km north of the wellfield.

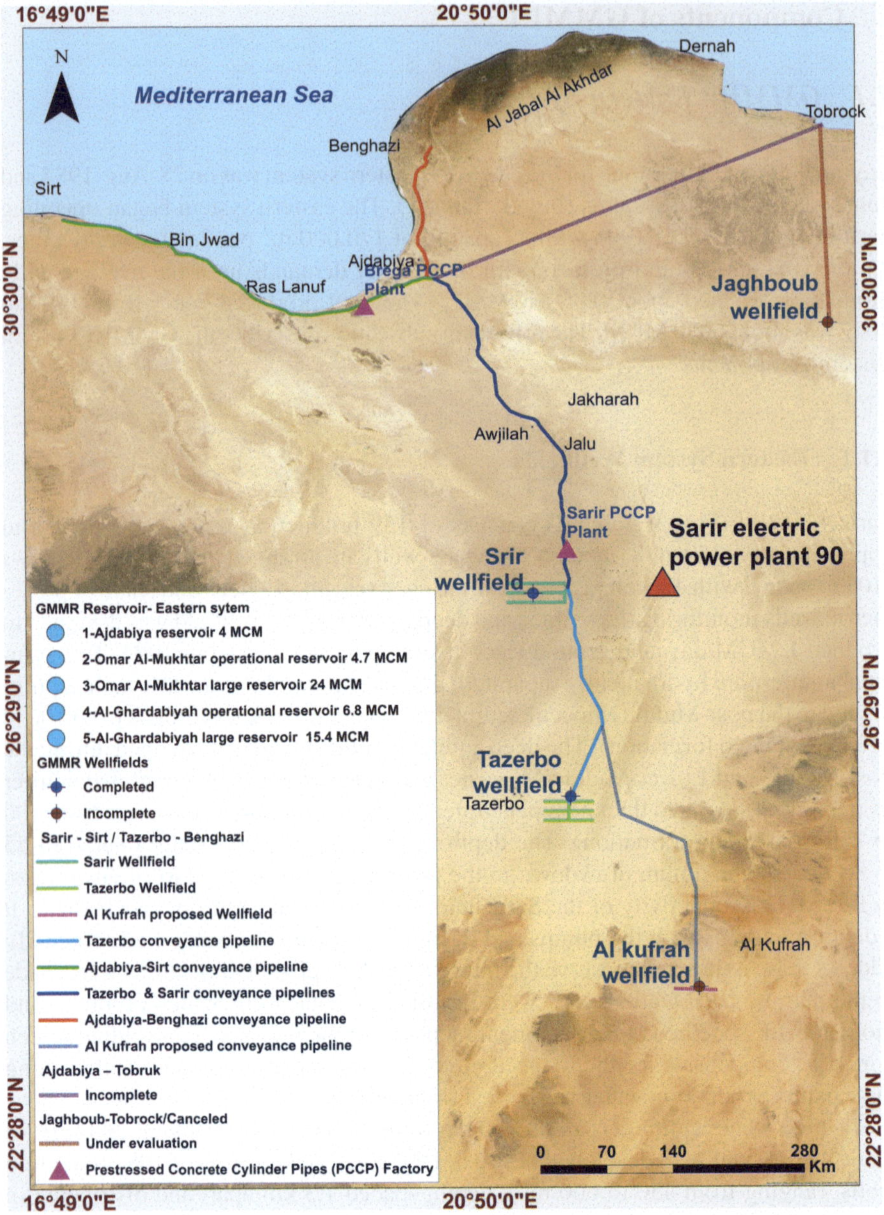

Fig. 3 The components of the GMMRP eastern system map based on data from GMMRA (2022)

Table 1 GMMRP eastern system (GMMRA 2022)

	Tazerbo-Sarir-Sirt-Benghazi	Al Kufrah-Tazerbo	Ajdabiya-Tobruk	Jaghboub-Tobrock
Designed transport volume m^3/day	2,000,000	1,650,000	Connection	–
Urban m^3/day	950,000	0	700,000	–
Agricultural m^3/day	1,000,000	1,500,000	0	–
Industrial m^3/day	50,000	0	10,000	–
Completion %	100	70	Preliminary studies	Cancelled

Table 2 Water wellfields of the GMMRP eastern system (GMMRA 2022)

		Sarir	Tazerbo	Al Kufrah
No. production wells		159	108	300
No. piezometric wells		12	35	14
Well depth (meter)		436–478	440–660	400–600
Single water well yield (litre/sec)		75	120	75
Designed production MCM/daily		1	1	1.68
TDS mg/L	Min	800	181.00	172
	Max	1000	229.45	704
Year of production		1993	2004	Not implemented
Total production from start to 2022 MCM/year		4289	2021	0
Current production Sarir and Tazerbo m^3/day		850		0

The water is abstracted from the deep aquifer (Upper Silurian Acacus formation), consisting of unconsolidated medium to very coarse-grained sandstone containing quartz and iron oxide with clay and siltstone components. The aquifer is overlaid by the aquitard (Lower Devonian Tadrart formation) and under-laid by the aquitard (Acacus Formation Lower–Upper Silurian). The depth of the aquifer ranges between 263 to 514 m. Aquifer thickness ranges from meters 105 to 192 m. The recorded drawdown in the piezometric since the start of abstraction is 22 m. Tazerbo water wellfield has relatively high concentrations of iron, manganese and carbon dioxide, and a purification system has been installed on the well heads to purify the water. Periodic monitoring of the aquifer's behavior showed that there is a good hydraulic connection in the region, especially between the intermediate and deep aquifers.

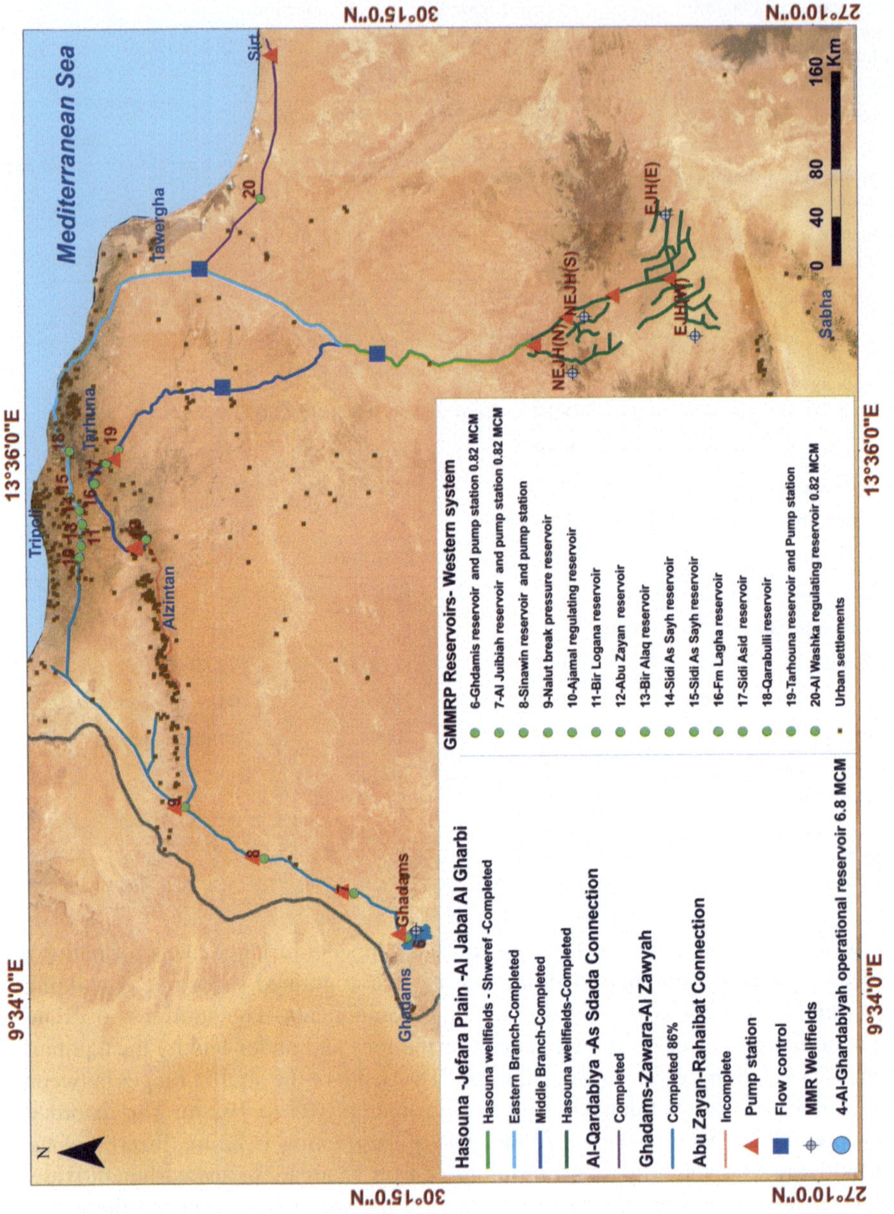

Fig. 4 The components of the GMMRP western system based on data from (GMMRA 2022)

2.1.2 Eastern System Conveyance Pipeline

The subsystem Sarir-Sirt-Tazerbo-Benghazi is 1586 km long and has a diameter between 3.6 and 4 m, and the collection pipeline in Sarir and Tazerbo wellfields is 339 km long. The subsystem represents phase I of GMMRP, which was implemented first, where it was designed to transport 2 MCM of freshwaters daily via two conveyance pipelines. One conveyance pipeline extends from the Tazerbo wellfields, and the other from the Sarir wellfields. Both wellfields have two collection tanks (Header tanks) with a total storage capacity of 510,000 m^3. Under gravity, both conveyance pipelines head north to collect in a balancing and regulating reservoir in southern Ajdabiya city, which has a capacity of 4 MCM. From the Ajdabiya reservoir, the conveyance pipeline is divided into two branches. The first branch of conveyance pipelines heads east to the Benghazi region to transfer 1.18 MCM as designed. It ends in the Omar Al-Mukhtar operational reservoir, which has a capacity of 4.7 MCM and is connected to the Omar Al-Mukhtar large reservoir (seasonal), which has a total of 24 MCM. The second branch of the conveyance pipelines heads west to transfer 820,000 m^3 to settle in the Al-Qardabiya operational reservoir south of Sirte city, with a capacity of 6.8 MCM connected to Al-Ghardabiyah large reservoir (seasonal), which has a total of 15.4 MCM.

Al Kufrah-Tazerbo connection aims to boost production capacity in the eastern system. The exploratory wells were drilled, the wellfield location near Al Kufrah oasis was identified, and 70% of the conveyance pipeline system (385 km) was installed. Furthermore, the Ajdabiya-Tobruk connection is still being engineered, and the proposed subsystem Jaghboub-Tobrock has been cancelled in light of the exploratory drilling results and the drilling project's cost.

2.2 *GMMRP Western System*

Laying the foundation stone for the GMMRP western system was on 26-Aug-1986. The phase consists of three subsystems (Fig. 4; Table 3). The components of the GMMRP western system are described in the following subsection.

Table 3 GMMRP western system (GMMRA 2022)

	Hasouna-Jefara plain	Al-Qardabiya-As Sdada connection	Ghadams-Zawara-Al Zawyah
Designed transport volume m^3/day	2,500,000	980,000	250,000
Urban m^3/day	1,200,000	–	250,000
Agricultural m^3/day	1,250,000	–	–
Industrial m^3/day	50,000	–	–
Completion %	100	100	86

Table 4 Water wellfields of the GMMRP western system (GMMRA 2022)

		Hasouna wellfields				Ghadams
		NEJH(S)	NEJH(N)	EJH(W)	EJH(E)	
No. production wells		90	78	160	156	106
No. piezometric wells		22				16
Well depth (meter)		390–600				660–740
Single water well yield (litre/sec)		45–65				60–40
Designed production MCM/daily		2.5				0.250
TDS mg/L	Min.	563	729	601	764	606
	Max.	1386.00	1973	1552	1599	1294
Year of production		1996				2009
Total production from start to 2022 MCM/year		6917				0.013

2.2.1 Western System Wellfields

Hasouna wellfields (Fig. 4; Table 4) are divided into four wellfields; North East Jabal Hasouna South (NEJH(S)), North East Jabal Hasouna North (NEJH(N)), East Jabal Hasouna West (EJH(W)) and East Jabal Hasouna East (EJH(E)). The wellfields consist of 476 production wells and 22 piezometric wells to monitor shallow and deep aquifers. Production water wells drilled to approximately (450–600) meters. The configurations of the Hasouna Wellfields were optimized to minimize the effects of water level drawdown in the wellfields area. The wellfield discharge was initially planned at a rate of 2.0 MCM/day and subsequently increased to 2.5 MCM/day. A water wells spacing of 1500 m was assumed in the optimizing scheme. Hasouna formation (Cambro-Ordovician) represents the primary aquifer exploited in the wellfields, consisting of sandstone with minor intercalation of clay and silt, and the thickness of the aquifer ranges from 300 to 800 m. The aquifer is unconfined in the wellfields' central and southeastern regions, while it is confined in the northern and northeastern wellfields areas. The depth to groundwater level ranges between 163 to 199 m. The recorded drawdown in the piezometric since the start of abstraction is 20 m at Pz well No.18 in the confined zone in the northern and northeastern wellfields and 8 m at Pz well No.11 in unconfined zone in the south.

Ghadams (Fig. 4; Table 4) consists of 106 production water wells only 52 wells drilled to a depth of approximately 660–740 m. The water wells are arranged in five parallel rows 3 km apart in addition to 16 piezometric wells. The wellfield was designed to produce 0.250 MCM/day, and groundwater abstraction commenced partialy in 2009. The aquifer system comprises the lower Cretaceous and upper Jurassic, composed of sands from Kiklah and Kabaw formations. Kiklah Formation consists of continental, unconsolidated to semi-consolidated sediments of Jurassic and early Cretaceous age which are composed of white to pink, fine to very coarse quarzitic sandstone and gravel with intercalations of silt, clay, shale and limestone. The thickness of the aquifer ranges from 300 to 800 m. Most of the water wells are artesian. The recorded drawdown in the piezometric since the start of abstraction is 2–5 m.

2.2.2 Western System Conveyance Pipeline

The conveyance pipeline for Hasouna Jefara Plain Al Jabal Al Gharbi is 983 km long and has a diameter between 3.6 and 4 m (Fig. 4; Table 4). The collection pipes of Hasouna wellfields with diameters ranging from 300 to 4 m and a total length of 753. Four pump stations are used to pump the water from the wellfields into the conveyance pipeline, which branches into two (the central and eastern) branches at the flow-control station in Shweref at a distance of about 170 km from the wellfields. The middle branch runs north until it reaches the Wadi Tamala flow-control station and then to the regulating reservoir and pump station in Tarhuna city and other regulating and break-pressure reservoirs. It passes to feed the urban settlements in the southern Jefara plain area. From there, water is pumped using two pump stations in the Al Yaaqeeb area to the Abu Zayan reservoir (Abu Zayan-Rahibat connection) to supply the Al Jabal Al Gharbi urban settlements. The eastern branch heads to the northeast until the turnout at the Al-Sadda pumping station to the Al-Washka regulating reservoir with a capacity of 0.82 MCM meters and leads to the Al-Qardabiya reservoir to connect with GMMRP eastern system. While the main conveyance pipeline turns to the west parallel to the Mediterranean Sea coast to supply the urban settlements in the Jefara plain.

The conveyance pipeline Ghadams-Zawara-Al Zawyah is 821 km long and has a diameter between 3.6 and 4 m (Fig. 4; Table 4). The collection pipes of Ghadams wellfield with diameters ranging from 300 to 4 m and a total length of 753 km. The collected water is pumped from the wellfield to Ghadmis reservoir 0.82 MCM and pumped in two stages, the first to the Al Juibiah reservoir and pump station and the second to the Sinawin reservoir and pump station that boosts it to the Nalut brake pressure reservoir. From there, the conveyance pipeline branches to Al Harabah and the western parts of Al Jabal Al Gharbi, the southwestern regions of Jefara plain, and heads to the north to Al Zwayah and Zawarah cities.

3 Management and Implementation of GMMRP

The Great Man-made River Project Implementation and Management Authority (GMMRA) was established in Benghazi city to implement and manage the GMMRP under the Council of Ministers (formerly General People's Committee GPC) after the decision to finance and implement the GMMRP was made in October 1983, as affected by law No (11), and in the same year. The mandate of GMMRA included planning, design, implementation, operation, maintenance, and administration. As a result, the GMMRA was established as a public entity with administrative and financial independence. And so was the financing and contracting for implementation were governed by several organized laws and decrees (Table 5) issued mainly by the GPC.

To ensure the availability of sufficient and continuous funding for the GMMRP implementation, operation and maintenance, Law No. 10 (1983) was issued. Then several complementary laws were given, including Law No. 5 (1986), Law 19 (1991) and Law 21 (1994), as shown in Table 7. According to these laws, the GMMRP is funded through a range of sources, including the general budget, loans, water sale fees, revenues from GMMRA implemented projects or facilities, fixed and movable assets, donations and gifts, taxes on tobacco, fuel products, money transfers, tickets, and license renewal. Table 6 shows the investment cost to date for the GMMRP.

The financial resources required to complete the remaining phases of the GMMRP were estimated at 4.088 billion USD$ as in Table 7.

The GPC Law No. (218/1993) decree authorized the Phase I GMMRA water price (Table 8). A detailed study on the costs and pricing of the GMMRP's water for phases I and II was later prepared by the GMMRA. Based on the findings of the study, a

Table 5 Main laws and decrees govern the GMMRP

No.	Type	Year	Subject	Citation
10	Law	1983	Financing and implementation of the MMRP	GPC (1983a)
11	Law	1983	Establishment of the GMMRP Execution and Management Authority (GMMRA)	GPC (1983b)
5	Law	1986	Amending some provisions of Law No. 10 of 1983 regarding the financing and implementation of the GMMRP	GPC (1986)
89	Decree	1984	Approving the financial regulations for the implementation and management of the GMMRP	GPC (1984a)
90	Decree	1984	Approving the contracting regulations regarding GMMRA, organizing bids and contracting	GPC (1984b)
91	Decree	1984	Approving the training and development regulation for the GMMRA	GPC (1984c)
92	Decree	1984	Approval of the procurement regulations for the GMMRA	GPC (1984d)
19	Law	1991	Concerning the approval of some provisions related to the GMMRP and the investment in its waters	GPC (1991)
21	Law	1994	Amending a provision in Law No. 19 of 1991 establishing some provisions related to the MMRP and the investment of its waters	GPC (1994)
18	Law	1996	It is amending the value of some financial fees prescribed for Financing the GMMRP, investing its waters, and deciding some provisions	GPC (1996)
626	Decree	2008	Permission for the GMMRA to participate in establishing a joint venture company	GPC (2008)

Table 6 Detailed investment cost of the GMMRP (values in USD $1,000,000's based on the exchange rate from 2013 (i.e. 1 USD$ = 1.25 Libyan Dinar (UNICEF 2019; GMMRA 2013)

No.	Item	Tazerbo-Sarir-Sirt-Benghazi	Al Kufrah-Tazerbo	Al-Qardabiya-As Sdada	Hasouna-Jefara plain	Ghadams-Zawara-Al Zawyah
1	Investment expenditures on contracts	949	746	449	751	1202
2	Assets	1415	0	0	2437	0
3	Loan burdens	241	0	0	216	0
4	Running expenses	265	0	0	236	88
5	Assets to operate systems	33	0	0	22	0
Subtotal		2903	746	449	3662	1290
Total						9050

Table 7 Required financial resources to complete the GMMRP phases

Item	Value in USD$ 1000,000's
Completion of the existing phases	2811
New projects	894
Operation and maintenance	383
Total	4088

proposal to modify the current pricing was made; it was approved in 2008 (Table 8), but it was never actually made public. Urban and Industrial water uses is paid by the General Company for Water and Wastewater (GCWW) from the GMMRA.

The GMMRA has a comprehensive organizational structure that has been approved (Fig. 5; Table 9). The three semi-independent branches of the GMMRA are headquartered in Benghazi (Eastern system) and Ben Ghashir near Tripoli (Western system). A management council of five members governs the GMMRA. A career staff guide was prepared. Training, salaries, health, safety and environment guides also exist. Separate organizational structures exist for each system (GMMRA 2022; UNICEF 2019). From planning to design to implementation to operation and maintenance, the GMMRA's work scope has evolved. That is, the GMMRA's primary focus shifted from construction to operation and maintenance, which resulted in a parallel shift in the number and specializations of human resources, with increases and decreases in line with changes. This calls for ongoing training and qualification of human resources to keep up with change. The GMMRA was able to understand this and respond to it with ease, as shown by their competency in operating and maintaining GMMRP components without needing to replace them.

The MMRA operates with a workforce of 2472 people, with 70% working in technical positions due to the nature of the GMMRP. Administrative staff comprise 27% of the workforce, while financial and legal teams make up 2.5 and 0.5%, respectively. The headquarters and eastern systems employ most GMMRP personnel (33 and 32%, respectively); the western system employs 27% of the total. The GMMRP's high-level performance over its operational life suggests that the staff qualifications and geographic distribution are appropriate. The nature and size of the human resources will change as the project's functions shift from construction to operation and maintenance. This shift is expected to be easily absorbed by the existing workforce.

Table 8 The current and proposed GMMRP water pricing (UNICEF 2019; GMMRA 2013)

Water use	Current price in USD$ for the cubic meter of water	Proposed pricing in USD$ for the cubic meter of water
Urban	0.064	0.096
Agriculture	0.038	0.050
Industrial	0.637	0.640

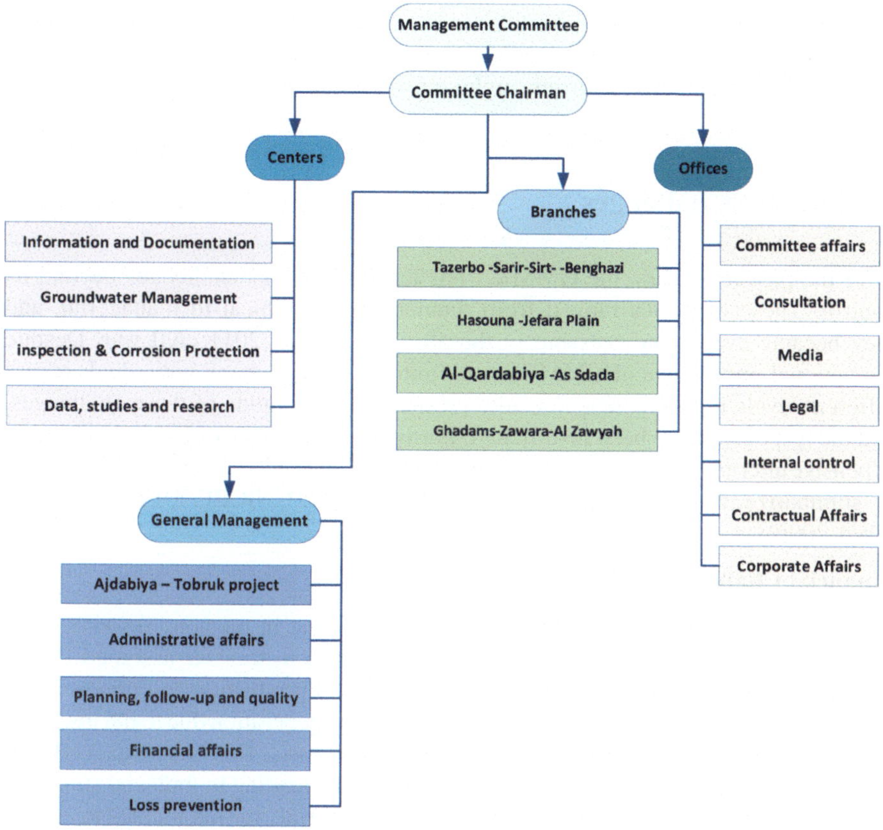

Fig. 5 Organizational structure of GMMRA (2022)

Table 9 The workforce of the GMMRA (2022)

Administrative divisions	Management	Technical	Financial	Legal	Total
General management, offices and centers	425	306	56	13	800
Tazerbo-Sarir-Sirt-Benghazi	82	736	1	0	819
Hasouna-Jefara plain	85	586	4	0	675
Al-Qardabiya-As Sdada	13	33	0	0	46
Ghadams-Zawara-Al Zawyah	58	74	0	0	132
Total	663	1735	61	13	2472

The GMMRA's organizational structure includes an information and documentation center, a data, studies, and research center, as well as an information technology unit, which indicates the organization's primary interests in information and data. The training facility is in charge of adapting the information center staff to the evolving information system requirements. All work operations are assigned numbers, and technical, financial, and administrative reports are created regularly and upon request using the data kept in the various GMMRA units. A central network links the different GMMRA units with this electronic archiving system.

4 Current Challenges

Over the past few years, the GMMRA has faced numerous challenges beyond its control. These obstacles ranged from financial and technical to managerial, and they became even more pressing in the aftermath of the 2011 civil war. Despite the national importance of GMMRP as a strategic project, many individuals across different levels in Libya have not fully grasped the magnitude of these challenges, and the responses have been insufficient. Furthermore, authorities have sought more expensive alternatives to provide the regions where GMMRP is meant to be delivered. To summarize, the challenges include (GMMRA 2022; UNICEF 2019; NPC 2014; GWA 2006):

Financial Challenges

- **Decline in Financial Resources**: Significant decline in financial resources due to the failure of the Central Bank of Libya (CBL) to transfer assigned tariffs, political conflicts, decreased imports, and lower oil production and prices.
- **Revenue and Cost Disparity**: Insufficient revenue generation due to the low price charged to consumers compared to high production costs. This issue is exacerbated by the GCWW's failure to pay water fees promptly, leading to financial shortfalls and inability to import crucial spare parts and materials for maintenance.
- **Implementation Costs**: Medium-term development investment plans cannot be accurately estimated without comprehensive information, and the estimated costs exceed GMMRA's current financial capability.
- **Electricity and Loan Burdens**: High costs of electricity consumption and financial burdens from loans taken to repay letters of credit.
- **Exchange Rate Fluctuations**: Increased implementation contract prices due to fluctuations in the exchange rate and the increased value of the dollar compared to the Libyan dinar.
- **Institutional Instability**: Frequent changes in the institutional dependency for the GMMRA post-2011 have impacted internal management.

Technical Challenges

- **Infrastructure Damage**: Significant direct and indirect financial losses from armed conflict, including damage to components, unauthorized connections, equipment replacement, and maintenance.
- **Water Supply Interruptions**: Increased frequency and duration of water supply interruptions due to frequent electricity supply interruptions.
- **Vulnerability to Pollution**: Groundwater basins are vulnerable to pollution from human activities, especially oil production.

- **Measurement and Data Gaps**: Lack of precise measurement of water supplied to consumers and disruption in information flow due to attacks on the acoustic monitoring system.
- **Illegal Connections and Leaks**: Damage to control rooms, valves, and pipes from illegal connections, causing significant water losses.
- **Aging Infrastructure**: Issues with conveyance pipelines and water tanks, including leaks after 25 years of operation, compounded by a lack of skilled workers and limitations in engaging foreign specialists due to security concerns.

Managerial Challenges

- **Water Allocation Criteria**: Unclear criteria for water allocation among various users, necessitating clarification and alignment with economic, technical, social, and environmental standards.
- **Component Failure Impact**: The interconnected nature of GMMRP infrastructure means that failure in one segment causes service interruptions in subsequent segments.
- **Remote Infrastructure Protection**: The vast and remote infrastructure is difficult to protect, and specialized equipment is challenging to replace or acquire locally.
- **Corrosion Issues**: Possible corrosion of prestressed concrete pipes due to damage to cathodic protection systems and theft of components.

5 Conclusion and Recommendations

The stated goal at the launch of the MMRP is to overcome the scarcity of water in the cities of the Libyan north, especially with population increase and water resources scarcity. The project aims to supply the cities of the north with available groundwater from the southern basins through water wells and a vast conveyance pipeline system. There is still controversy over the viability of the GMMRP, primarily as it depends on non-renewable groundwater. However, about 25 years have passed since the operation of the GMMRP, during which the project provided a viable solution to the problem of water scarcity in northern Libya. Thus, it achieved its objectives and confirmed its technical, economic and social feasibility without adverse environmental impacts. Water production has not exceeded half the designed production capacity. The groundwater level drawdown is still beyond the levels indicated in the mathematical models implemented during the GMMRP feasibility studies during the 1980s. At the same time, the employees of the GMMRA have acquired distinguished experiences and accumulated large amounts of data, information and lessons learned in planning, design, implementation and operation during the various stages of project development, which are unique due to the nature of the project and thus a reference base for workers and those interested in water transport at a significant local and global level. The project can also meet the urban water needs of most of the Libyan population for several decades. Finally, the GMMRP is a strategic and promising solution for Libya's water scarcity. Nevertheless, the challenges associated with the project demand prompt action to address them and ensure the project's sustainability. Accordingly, it recommends the following:

Financial Recommendations

- **Develop a Sustainable Financial Plan**:

  Diversify revenue streams, reduce costs, and improve collection efficiency. Implement a financial management system to monitor and track revenues and expenditures.

- **Implement Financing Regulations**:

  Ensure compliance with all GMMRP financing regulations. Engage in dialogue with the CBL to address exchange rate issues and seek alternative funding sources for importing necessary parts and supplies.

- **Establish Partnerships and Collaborations**:

  Secure funding, expertise, and resources by partnering with other Libyan organizations and international donors.

- **Examine Payment Plans**:

  Ensure prompt and complete payment of water fees by GCWW and consider government subsidies to lower production and consumption costs.

- **Improve Billing and Collection**:

  Deploy advanced billing technologies and regularly use flow meters for accurate invoicing and consumption tracking.

- **Conduct Feasibility Studies**:

  Accurately estimate the cost of medium-term development investment plans, assessing their feasibility and financing options.

Technical Recommendations

1. **Prioritize Infrastructure Maintenance**:
   - Implement a comprehensive maintenance strategy, including regular inspections, preventive maintenance, and timely repairs.
2. **Install Water Measurement Devices**:
   - Ensure accurate measurement of water consumption at all distribution points to enhance data collection and analysis.
3. **Implement Pollution Control Measures**:
   - Protect groundwater basins from contamination, particularly from oil production activities.

4. **Improve Water Supply Reliability**:
   - Enhance electricity supply reliability and explore alternative energy sources such as solar power.

Managerial Recommendations

1. **Develop Clear Water Allocation Criteria**:
   - Establish transparent criteria for water allocation based on economic, technical, social, and environmental standards.
2. **Strengthen Security Measures**:
   - Prevent illegal connections and reduce water losses through increased surveillance and regular infrastructure maintenance.
3. **Reduce Water Waste**:
   - Promote water-saving technologies and practices to encourage responsible usage and reduce production costs.
4. **Develop Effective Monitoring and Inspection Systems**:
   - Regularly inspect pipelines and accessories, and invest in advanced monitoring and surveillance technologies.
5. **Ensure a Stable Institutional Framework**:
   - Maintain a stable governance structure for long-term stability and continuity.

Strategic Recommendations

1. **Integrate Planning for Irrigation Water Management**:
   - Invest in treated wastewater to recharge depleted coastal aquifers and preserve high-quality groundwater.
2. **Establish a Comprehensive National Water Strategy**:
   - Prioritize water supplies for urban, industrial, and agricultural uses, considering all relevant factors.
3. **Give GMMRP National Priority**:
   - Prepare and implement annual plans to ensure the operation and maintenance requirements of GMMRP.
4. **Evaluate GMMRA Performance**:
   - Identify gaps, strengths, challenges, and lessons learned to improve GMMRA's performance.
5. **Develop Alternative Water Resources**:

- Prepare for emergencies by developing alternative water resources to use during interruptions.

6. **Emergency Preparedness Plan**:
   - Include GMMRP recommendations in national strategic plans to handle long-term emergency outages.
7. **Capacity-Building Programs**:
   - Build a skilled workforce for infrastructure maintenance and repair, exploring partnerships with foreign firms if necessary.
8. **Raise Awareness**:
   - Ensure public awareness of GMMRP's viability and importance as a national strategic project, promoting its protection and sustainable use

References

Bakhbakhi M, Salem O (1999) "Why the Great Man-Made river project?" In: General Water Authority, Jamahiriya LA (eds) Regional aquifer systems in arid zones—managing non-renewable resources. UNESCO, Tripoli, pp 3–16

BCS (2006) Libya population and housing census 2006. Libyan Bureau of Census and Statistics

BCS (2020) Estimated Libyan population by regions for the year (2020). Libyan Bureau of Census and Statistics, Tripoli

Binsariti A, Saeed FS (2000) Groundwater salinity variations in the Cambro-Ordovician aquifer of eastern Jabal al Hasawnah, the Great Man-Made River Project, Libya. In: Sola MA, Worsley D (eds) Geological exploration in Murzuq Basin. Elsevier, pp 1–16. https://doi.org/10.1016/B978-044450611-5/50003-9

Gijsbers PJA, Loucks DP (1999) Libya's choices: desalination or the Great Man-Made river project. Phys Chem Earth Part B 24(4):385–389. https://doi.org/10.1016/S1464-1909(99)00017-9

GMMRA (2013) Status of GMMRP status and future plans. Benghazi

GMMRA (2022) "Great Man-Made River Project Authority." Retrieved 6 Oct 2022. https://gmra.com.ly/index.php/en

GPC (1983a) Law 10 of 1983 financing and implementation of the Great Man-Made river project. General People's Committee

GPC (1983b) Law 11 of 1983 establishment of the Great Man-Made river project execution and management authority (GMMRA). General People's Committee, Libya

GPC (1984a) Decree 89 of 1984 approving the financial regulations for the implementation and management of the MMRP. General People's Committee

GPC (1984b) Decree 90 of 1984 approving the contracting regulations for construction for the implementation and management of the MMRP, organizing bids and contracting. General People's Committee

GPC (1984c) Decree 91 of 1984 approving the training and capacity development regulation for the MMRA. General People's Committee

GPC (1984d) Decree 92 of 1984 approval of the procurement regulations for the MMRA. General People's Committee

GPC (1986) Law 5 of 1986 amending some provisions of law no. 10 of 1983 regarding the financing and implementation of the Great Man-Made River project. General People's Committee

GPC (1991) Law 19 of 1991 concerning the approval of some provisions related to the MMRP, and the investment in its waters. General People's Committee
GPC (1994) Law 21 of 1994 amending a provision in law no. 19 of 1991 establishing some provisions related to the MMRP and the investment of its waters. General People's Committee
GPC (1996) Law 18 of 1996 amending the value of some financial fees prescribed for financing the MMRP, investing its waters, and deciding some provisions in this regard. General People's Committee
GPC (1999) Great Jamahiriya's water situation and the national strategy for water resource management, Tripoli
GPC (2008) Decree 626 of 2008 permission for the execution and management authority of the MMRP to participate in establishing a joint company. General People's Committee
Guendouz A, Moulla AS (2010) The shared resources in the North-Western Sahara Aquifer System (Algeria-Tunisia-Libya): the use of environmental isotopes (Algeria Part). In: International conference "Transboundary aquifers: challenges and new directions"(ISARM2010), UNESCO, Paris, pp 8–11
GWA (2006) Water resources status of Libya, General Water Authority. Tripoli
Hamad S (2012a) Opportunities and challenges for groundwater artificial recharge by surface water harvesting in Libya. Libyan Agric Res Center J Int 3(6):260–268. https://doi.org/10.5829/idosi.larcji.2012.3.6.1112
Hamad S (2012b) Status of groundwater resource of Al Jabal Al Akhdar Region, North East Libya. 1(1):11
Hamad S, Younes A (2020) Evaluation of the Nubian sandstone aquifer system (NSAS) in Al Kufra Oasis, Southeast Libya. Desalin Water Treat. https://doi.org/10.5004/dwt.2020.25503
Hamad S, El Hasia A (2022) An overview of karst groundwater springs in Al Jabal Al Akhdar Region (North East Libya). Groundwater Sustain Livelihoods Equitable Growth (March):179–194. https://doi.org/10.1201/9781003024101-10.
Jones JR (1971) Ground water provinces in Libyan Arab Republic. In: Geology of Libya, Faculty of Science, University of Libya, Tripoli, pp 449–457
Mcloughlin PFM (1991) Libya's Great Manmade river project prospects and problems. Nat Res Forum 15(3):10–13. https://doi.org/10.1111/j.1477-8947.1991.tb00136.x
NPC (2014) National water security strategy. Tripoli
Pallas (1980) Water resources of the socialist people's Libyan Arab Jamahiriya. The geology of Libya. In: Second symposium on the geology of Libya, held at Tripoli, September (12–16) 1978. Academic Press, Tripoli, pp 539–593
Pim RH, Binsariti A (1994) The Libyan Great Man-Made river project. Paper 2. The water resource. Proc—ICE: Water, Maritime Energy 106(2):123–145. https://doi.org/10.1680/iwtme.1994.26389
Salem O (1992) Libya's future water supply the Great Manmade river project a partial solution to Libya's future water supply. Int J Water Resour Dev 8:37–41. https://doi.org/10.1080/07900629208722564
Shaki A (2006) Simulation and optimization studies for better groundwater management in the Murzuq basin, Libya. Heriot-Watt University
Siala TF, Stoner JR (2006) The Great Man-Made river project. In: Pipelines 2006: service to the owner, pp 1–8. ASCE Library
Swailem FM, Hamza MS, Aly AIM (1984) Isotopic composition of groundwaters from Kufra (Libya) as indicator for groundwater formation. Isotopenpraxis Isotopes Environ Health Stud 20(2):64–68. https://doi.org/10.1080/10256018408544979
UNECA (2005) Libya national water development report. Addis Ababa
UNICEF (2019) Assessment of water supply systems and institutions in Libya. Tripoli
Van Everdingen RO (1962) The deeper ground water in Libya. Int Asso Scient Hydrol Bull 7(3):33–39. https://doi.org/10.1080/02626666209493266
Wheida E (2012) The water resources management in Libya. Libyan Agric Res Center J Int 3(3):144–154. https://doi.org/10.5829/idosi.larcji.2012.3.3.1209

Mr. Salah Hamad is an assistant professor of water resources and geoinformation science at the Department of Natural Resources, Faculty of Natural Resources and Environmental Science, University of Omer Al-Mukhtar, Al Bayda, Libya. He holds an M.Sc. in Geographic Information Systems and Science from Salzburg University, Salzburg, Austria, in 2008. Mr. Hamad has been actively teaching undergraduate courses and supervising bachelor's degree students' graduation since 2010. His research interests lie in GIS mapping and spatial analysis, with a particular focus on geoinformatics research, geodata documentation, and the application of various GIS and remote sensing techniques in conventional water resources management, engineering, and other related aspects of earth science.

Mr. Fawzi Saeid is the former Director of the Technical Center for Groundwater Management (TCGM) at the Great Man-Made River Project Authority (GMMRA) and currently is the member of GMMRP management committee, in Benghazi, Libya, he is a senior hydrogeologist with over 35 years of expertise in the drilling industry, geology, hydrogeology, and geotechnical engineering. He obtained his M.Sc. in hydrogeology and groundwater management from the University of Technology, Sydney, Australia, 2004.

Groundwater: An Important Resource of Urban Water Supply in Libya

Salah Hamad and Rod Fensham

Abstract Libya is one of the world's poorest countries in renewable water resources, with a per capita share of water that is less than 10% of the global average. Water demand will rise with population, economic, and urban growth. Water use will also change in response to construction projects, security threats, climatic change, food security requirements, and crop demand increase, particularly in the north of the country where most of the population is concentrated. 97% of the country's total water consumption is sourced from groundwater. Springs, local water wells, and the Great Man-Made River Project (GMMRP) are the sources of groundwater used for urban water supplies. The springs slightly contribute to urban water supply in sparsely populated areas, primarily suburbs and villages. Most local water wells are located in urban areas with no other water sources, particularly Al Jabal Al Akhdar, the central regions, Jabal Nafusa, and the Jifarah plain. Only 53.7% of the 1591 water wells are operational. The GMMRP is currently the primary source of water supply for more than 70% of Libyan cities and will contribute to securing 2.3 billion m^3 of water annually if all phases are completed. The projected water demand, which is anticipated to be around 10 billion m^3 by 2035, cannot be met by the GMMRP alone. Complementary sources, like sewage treatment and saltwater desalination, must be developed to close the imbalance between supply and demand. Finally, water urban water supply services face serious challenges caused by ageing facilities, ongoing armed conflict, political, economic, and institutional instability, and continuous power and fuel supply cuts. Libya's water problem is more than just a quantitative shortage of fresh water.

Keywords Man Made river · Urban supply · Libya · Groundwater · Water scarcity

S. Hamad (✉)
Faculty of Natural Resources and Environmental Science, Omar Al Mukhtar University, Al Bayda 00218, Libya
e-mail: salah.hamad@omu.edu.ly

R. Fensham
School of Biological Sciences, Faculty of Science, University of Queensland, Brisbane, Australia
e-mail: rod.fensham@des.qld.gov.au

H. A. Zurqani (ed.), *Water Resources of Libya*, Springer Water,
https://doi.org/10.1007/978-3-031-80920-0_3

1 Introduction

Libya (Fig. 1) is one of the driest countries on earth. The climate is primarily arid, except for a narrow strip along the northern coast with a Mediterranean climate.

The northern part of the country receives 100–550 mm/year of rainfall, with the highest amounts falling in Al Jabal Al Akhdar in the northeast and Jabal Nafusa in the northwest (Fig. 2). Southern parts are hyper-arid, receiving less than 10 mm/year, forming a large portion of the Sahara Desert (NPC 2014; GWA 2006). The coastal belt is distinguished by relatively good soils and favourable climatic conditions, allowing for the development of agricultural, industrial development and densely populated urban areas (Bakhbakhi and Salem 1999; Pallas 1980).

The Libya population (Fig. 3) has grown from 1.56 million in 1964 to 6.9 million in 2020 (BCS 2020; Bakhbakhi and Salem 1999). However, 85% of the Libyan population occupies the urban center on the coastal strip, and the population growth is highest in these urban centers.

Around 50% of the world's population is expected to experience acute water shortages by 2030 as freshwater availability is expected to continue to decline (Aimar 2017). Both historically and currently, groundwater has been a critical water source for sustaining human life. Over the past century, usage of this resource has drastically increased (Fienen and Arshad 2016). Groundwater is the foundation of much of Africa's water supply, and its development is accelerating as the demand for safe water increases (Frédéric 2020). According to (MacDonald et al. 2012), groundwater storage in African aquifers is estimated at (0.36–1.75 million km^3). Although only a portion of this groundwater is available for extraction, the estimated volume is more than 100 times that of Africa's annual renewable freshwater resources (MacDonald et al. 2021). Moreover, groundwater is the only source of water supply for most of the local demand in North African arid/desert lands (agricultural, industry, tourism, and domestic) and, like the rest of the world, is undergoing water stress, defined as using more than 20% of their renewable water resources (Hamed et al. 2018). However, in the Middle East and North Africa (MENA) region, groundwater is an essential and dependable source of fresh water. Groundwater is the primary drinking and irrigation water source for many countries in this region. However, the MENA region faces the highest level of groundwater development pressures globally. Overexploitation, deteriorating quality, climate change, and a lack of transboundary aquifers threaten the reliability of this source of freshwater (WWAP 2015; Roudi-Fahimi et al. 2000; Aivar and Francois 2016).

Libya is one of the MENA countries where the demand for freshwater to sustain its economy, particularly its agricultural sector, exceeds traditional supplies. The increase in water consumption is out of proportion with the available capacity. Moreover, the scarcity of renewable water resources is a significant impediment to Libya's long-term development. Water consumption in Libya is rising, and the growth in water demand has markedly impacted Libya's water resources. Uncontrolled extraction has severely impacted the renewable northern coastal aquifers and has more

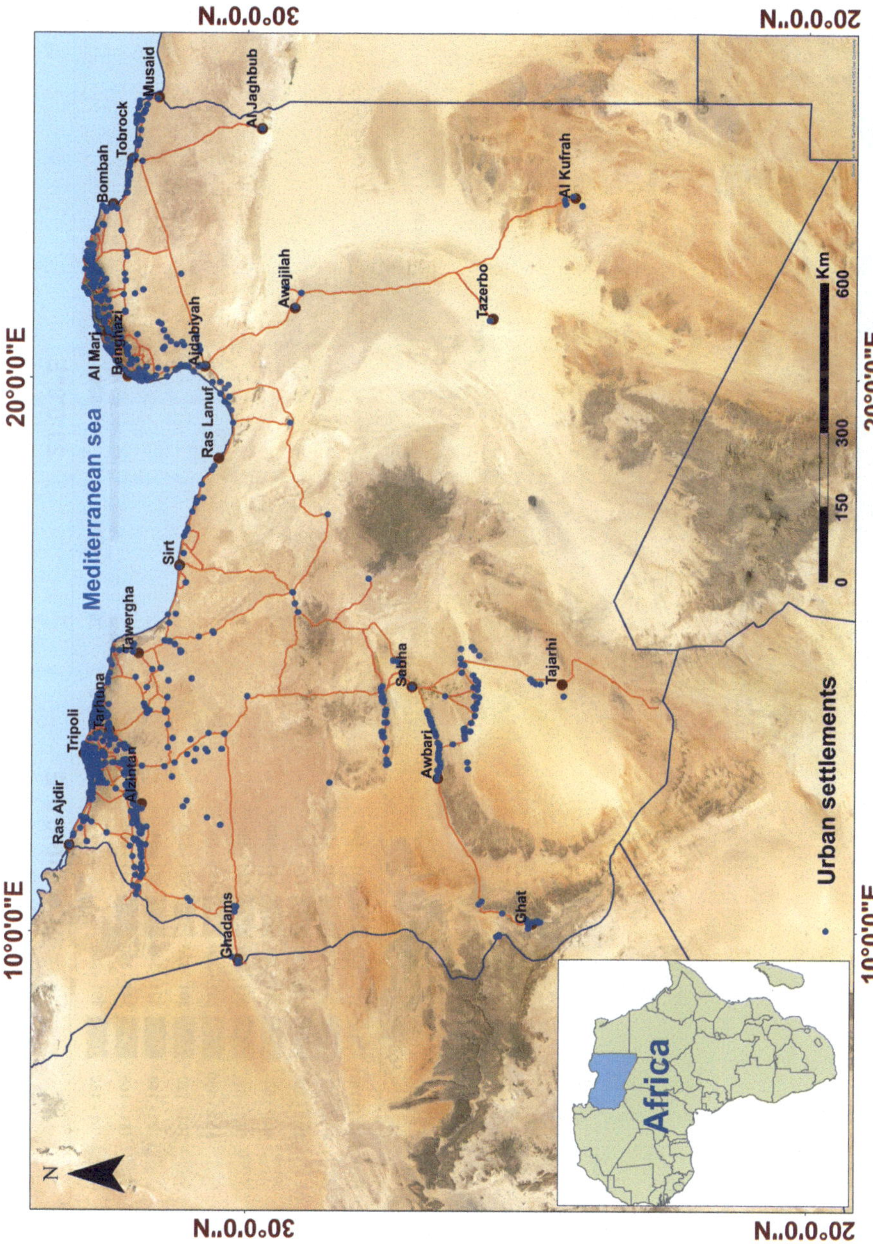

Fig. 1 Location map of Libya

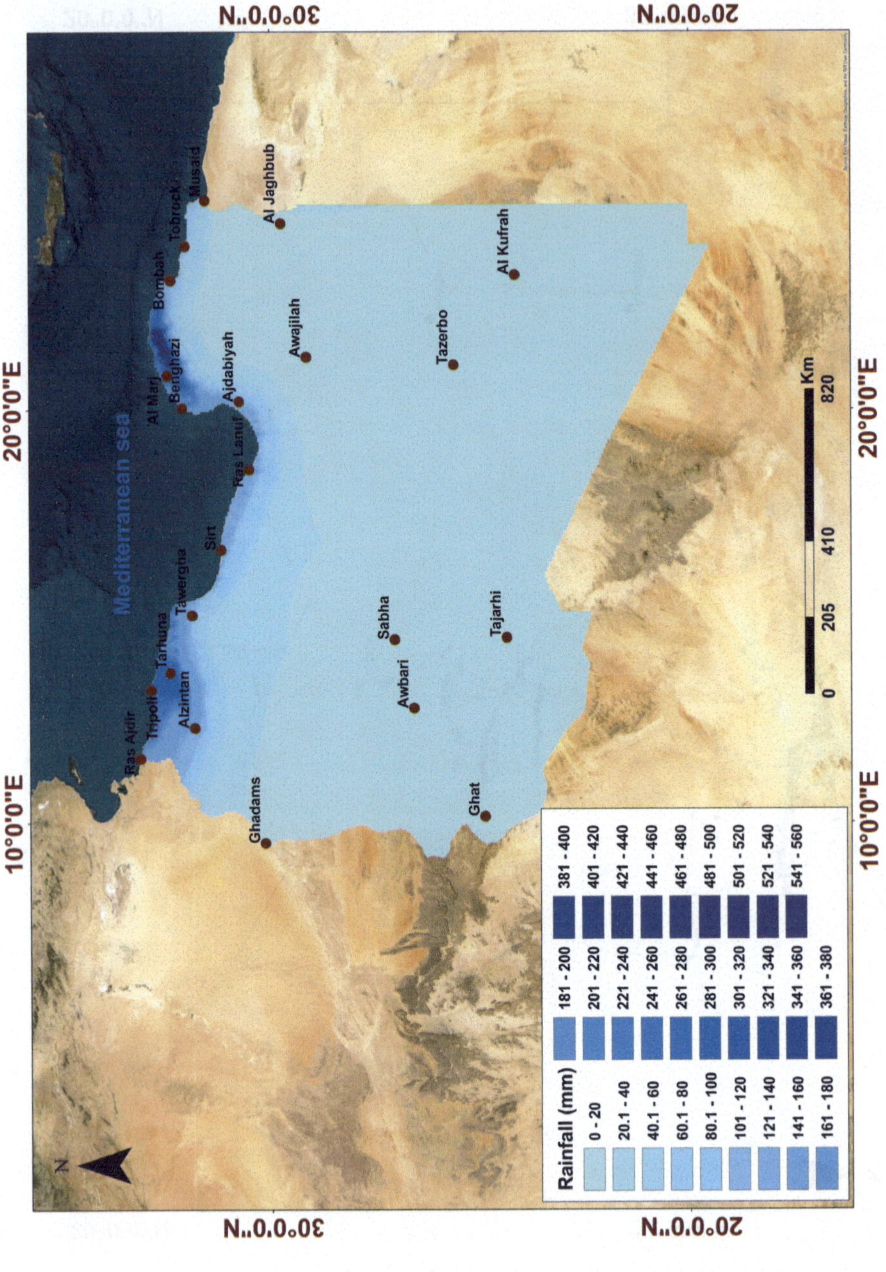

Fig. 2 Annual rainfall map of Libya (NPC 2014; GPC 1999; Pallas 1980)

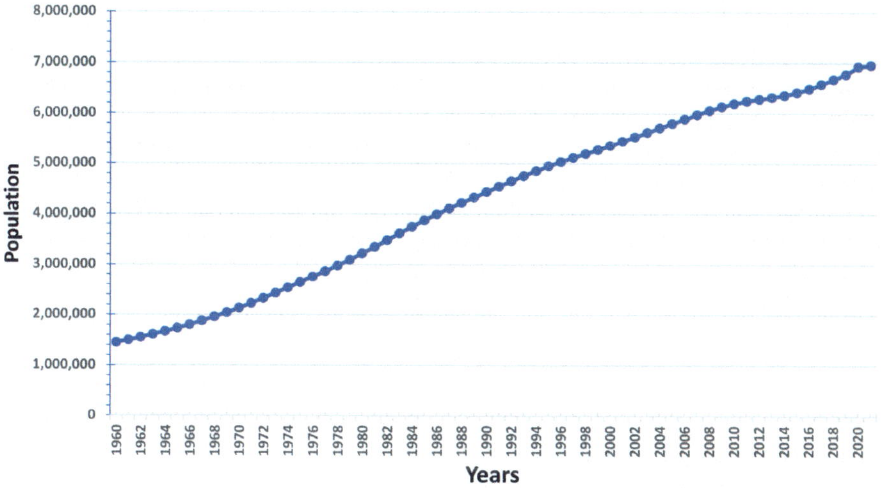

Fig. 3 Libya population (BCS 2020; World Bank 2020)

modest implications for the non-renewable southern aquifers (NPC 2014; CEDARE 2014; Hamad 2012; Alamin 2012; Abufayed and Elkebir 2010; GWA 2006).

The purpose of this chapter is to discuss the role of Libyan groundwater in providing urban water supply. Libya's geology and hydrogeology, as well as its water resource potential, are briefly reviewed. Also discussed is the current state of urban water supply, future projections, and the significant challenges that the Libyan state faces in combating water scarcity. Furthermore, the chapter concludes with recommended measures such as reducing the water deficit, protecting water resources from pollution, developing human and institutional capacities, and prioritizing the provision of GMMRP water for urban purposes, particularly in urban areas with acute water deficits.

2 Geology of Libya

Libya has four main geological basins: Sirt, Ghadames, Murzuq, and Al Kufrah, as well as the Cyrenaica platform onshore and two offshore basins: Benghazi and Sabratah (Fig. 4) Ghadamis, Murzuq, and Al Kufrah basins, are Paleozoic basins, whereas Sirt Basin is primarily a Mesozoic and Tertiary basin. The Ghadames Basin is located northwest of the country (Fig. 4) and continues westwards into Tunisia and Algeria. The Murzuq Basin is located southwest (Fig. 4) and continues southwards into Niger, where it terminates with the Tihimboka Arch. Al Kufrah Basin in the southeast extends into Egypt in the east and into Chad and Sudan in the south, although it has its maximum depth in Libya (Hallett and Clark-Lowes 2016; Selley

1997; Gouadrzi 1980; Klitzsch 1971). The Sirt Basin is situated entirely within Libya and extends northward offshore in the Gulf of Sirt.

Quaternary deposits generally mantle the coastal plains; and form dunes and gravel plains over one-third of the country (Gouadrzi 1970). Sedimentary rocks dominate the consolidated and unconsolidated surface geology of Libya except for a few areas the outcropped formations resulting from recent volcanic activity (Selley 1997), such as Jabal Nafusa in the northwest, Jabal As Sawdah and Jabal Al Haruj in the middle of the country, and Waw a Namus in the south. Palaeozoic rocks outcrop et al. Gargaf Arch, Jabal Akakus in the country's southwest, Dur al Qusah, Tibisti, Jabal Eghi in the southeast of Libya, Jabal Zalmah, Jabal Arknu, and Jabal Awaynat in the southeast part of Libya.

3 Hydrogeology of Libya

Libya can be divided into the northern division, consisting of aquifers where current rainfall provides sufficient recharge, and the southern division, considered to be non-renewable, receiving little or no recharge. (UNECA 2005; Bakhbakhi and Salem 1999; Pallas 1980). The north division includes the coastal plains (Figs. 5 and 6); Jifarah plain, Al Khomes Misrata-Tawergha plain, Sirt plains, Ajdabiya-Benghazi plains, and the narrow plains between the Mediterranean Sea and Al Jabal Al Akhdar, Dafna plateau, and Al-Batnan plateau. The mountain ranges of Al Jabal Al Akhdar in the northeast and Al Jabal Al Gharbi (Jabal Nafusa) in the northwest are essential to recharge areas, although rainfall, even in these areas, is mostly less than 500 mm per annum (FAO 2009; El-Tantawi 2005).

Groundwater in the northern division occurs in Quaternary, Tertiary, Upper Cretaceous, and Triassic aquifers. These aquifers are recharged by current rainfall (Pallas 1980). Still, the groundwater can be saline, especially in the coastal areas in the central part of Libya, as well as the Tobrock coastal area that extends from the Gulf of Bambah to the Egyptian border. The Al Jabal Al Akhdar, the Benghazi plain, and the Jifarah plain have high-quality groundwater, especially in sub-coastal areas, due to adequate recharge (Wheida 2012; Hamad 2012; Salem 1992; Shaki 2006).

In the southern parts of Libya, groundwater exists in predominantly sandstone aquifers of Tertiary, Lower Cretaceous, Triassic, and Palaeozoic ages (Hamad and Ahweej 2020; Binsariti and Saeed 2000; Salem 1992; Pallas 1980). These aquifers represent a massive reserve of high-quality water with a salinity of less than 500 mg/litre. The waters are of late Pleistocene and Holocene age (Guendouz and Moulla 2010; Swailem et al. 1984), are not currently being recharged, and are therefore non-renewable.

Libya has five groundwater management divisions (Groundwater systems) (Fig. 6; Table 1) identified by Pallas (1980), Jones (1971); Jifarah plain, Gadamis-Al Hamadah, Murzuq, AI Jabal Al Akhdar, Sirt-Sarir and AI Kufrah. These groundwater divisions include areas that are not strictly geological basins but have been erroneously indicated as groundwater basins in some sources.

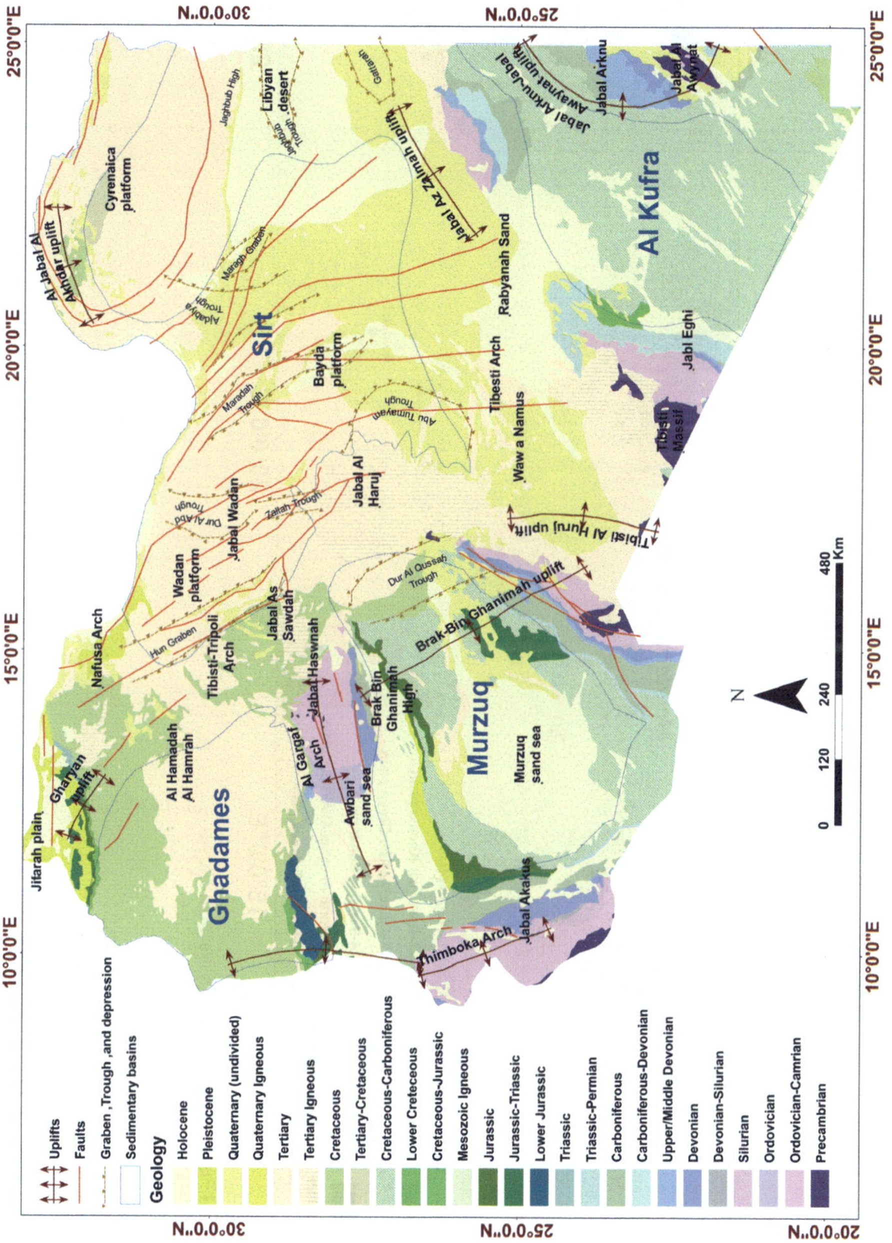

Fig. 4 The geology and structure of Libya modified after (Hallett and Clark-Lowes 2016; IRC 2009; Gouadrzi 1980)

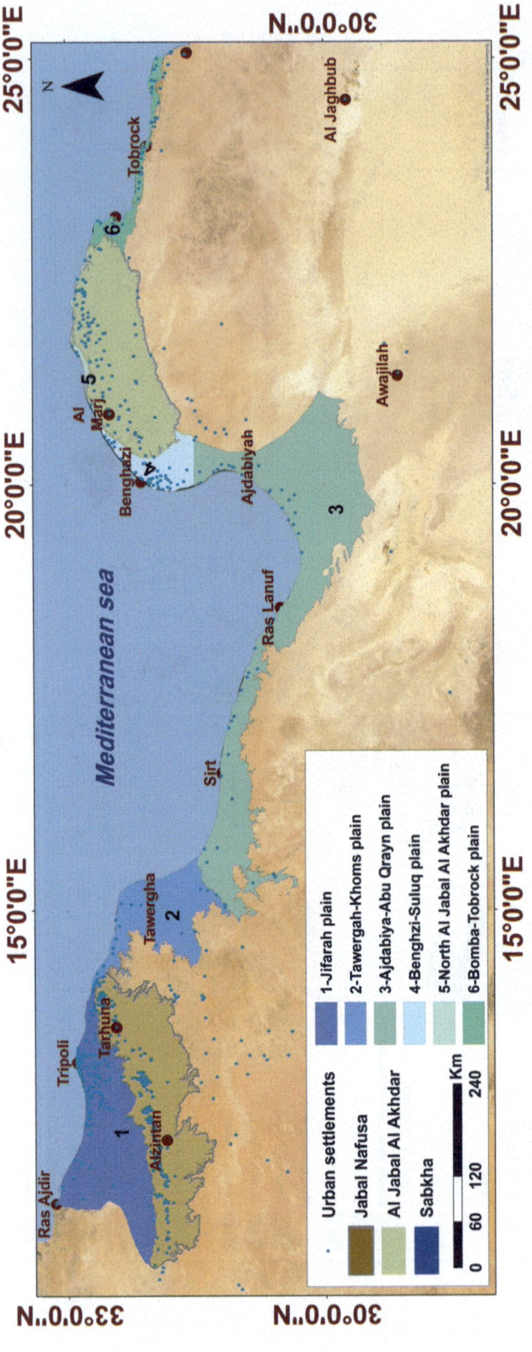

Fig. 5 Coastal plains and mountains in northern Libya

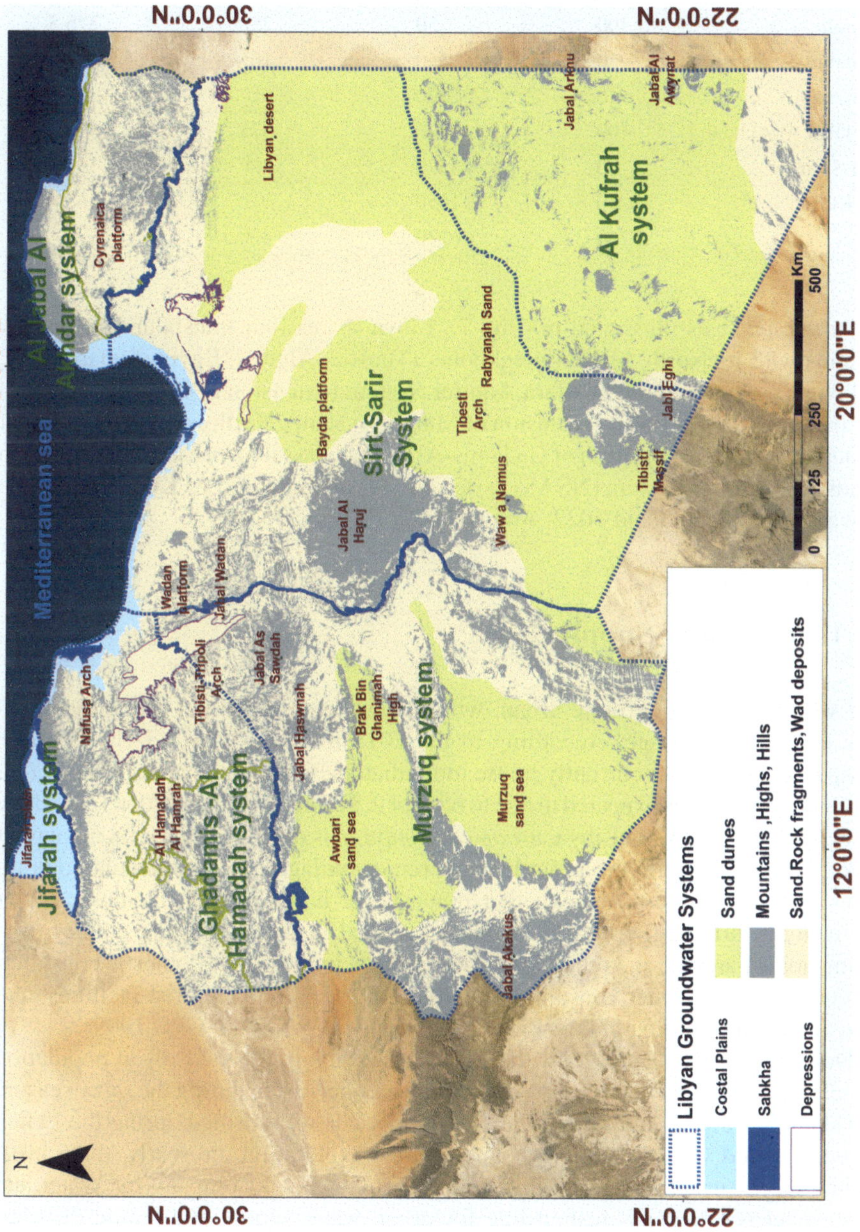

Fig. 6 Main physiographic features and groundwater systems modified after (Hamad 2020; FAO 2009; GPC 1999)

Table 1 Available conventional water resources in million cubic meter (MCM) (NPC 2014)

Groundwater systems	Groundwater Renewable	Groundwater Non-renewable	Surface water	Total
Jifarah	300	50	25.5	375.5
Gadamis-Al Hamadah	50	350	16.4	416.4
Murzuq	0	2500	0	2500
Al Jabal Al Akhdar	400	100	375	875
Sirt-Sarir	0	1570	3.4	1573.4
Al Kufrah	0	1650	0	1650
Total	750	6220	420.3	

Furthermore, as illustrated in (Fig. 7), Libya shares four transboundary aquifer systems with its North African neighbours (Tunisia, Algeria, Egypt, Sudan, Chad, and Niger), including the Djeffara Aquifer System (extension of the Jifarah system in Libya, Lake Chad Basin (extension of Murzuq system), Northwest Sahara Aquifer System (NWSAS) (extension of Gadamis-Al Hamadah system in Libya), and Nubian Sandstone Aquifer System (NSAS) (extension of Sirt-Sarir and Al Kufrah system in Libya) (Hamad and Saaid 2022; Ahmed 2020; Frédéric 2020).

4 Urban Water Supply

The world is becoming more urban, with human settlements and economic activities taking center stage. According to the 2018 revision of World Urbanization Prospects, while cities currently house more than half of the global population (4.2 billion people), this is expected to rise to 68% or 2.5 billion people by 2050 (Aboelnga et al. 2019; UN 2018). In Libya, *urban consumption* is defined as the total amount of water consumed within urban settlements (cities, villages, and residential communities) for various purposes, whether domestic or non-domestic. Domestic water use includes drinking, cooking, housing, personal hygiene, washing clothes, irrigating home gardens, and other domestic human activities. Non-domestic consumption comprises all water consumption within industrial units, tourist facilities, and recreational and commercial facilities (NPC 2014; GWA 2006; GPC 1999).

During the 1950s and 1960s, the urban water supply for the Libyan population came from natural springs, desert oases, rainwater harvested during the rainy season, and shallow wells. In addition, relatively deep wells were drilled during the Italian occupation and with the commencement of oil exploration. Presently, three institutions supply urban water to Libya's cities: The General Company for Water and Wastewater (GCWW) is responsible for urban water supply and public distribution networks. The GCWW manages supplied water from the wellfields managed by the Great Man-made River Authority (GMMRA), water wells operated by the GCWW, and desalinated seawater supplies operated by the General Desalination

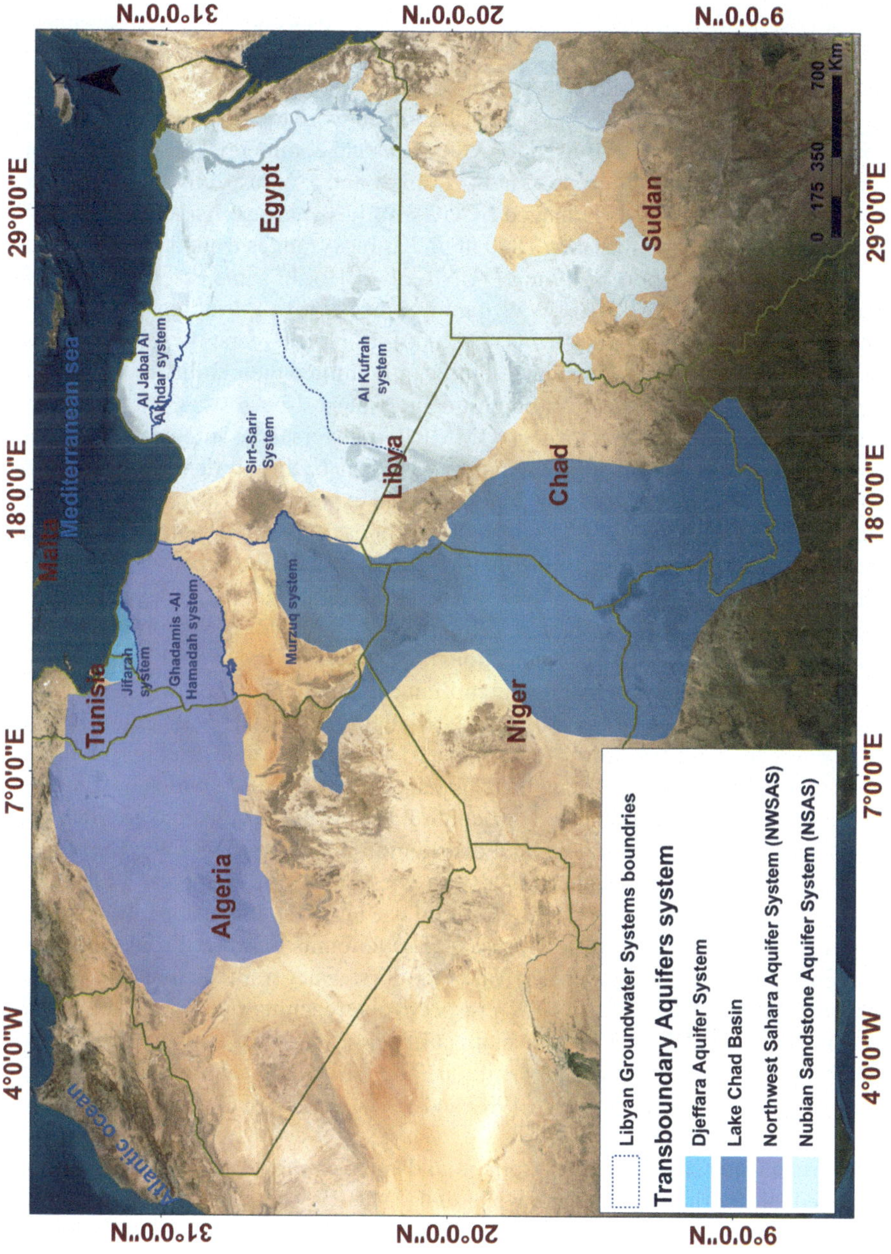

Fig. 7 Map of Libya's transboundary aquifers system modified after (IGRAC and UNESCO 2016)

Company (GDC) (NPC 2014). The GCWW buys water from the GMMRA and the GDC at officially set prices. At the same time, it collects fees for providing water and sanitation services from consumers. It receives support from the public treasury that covers the difference between water supply costs and collection revenues. Moreover, The GCWW manages 2964 water wells, 19 water treatment facilities, 2054 water tanks with a collective volume of 377,500 cubic meters, 806 pumping stations, transmission pipelines, and a distribution network totaling about 16,000 km in length. The supplied water is used for residential and commercial purposes and on a small scale for industrial purposes, at 80%, 15%, and 5%, respectively. The water quality from the water wells managed by GCWW has varying characteristics. The TDS ranges from 260 to more than 1840 mg/l, hardness ranges from 60 to 1050 mg/l, and chloride from 42.6 to 673.9 mg/l (UNICEF 2019). In some cases, the quality of the water supply from the GCWW may not meet Libyan drinking water regulations (NPC 2014; UNICEF 2019; GWA 2006). In recent years, these institutions have faced severe challenges that have significantly limited their ability to meet their obligations to provide adequate and high-quality water.

The Great Man-made River Project (GMMRP) transfers large amounts of groundwater from aquifers within Libya's main geological basins to densely populated coastal areas where agricultural development could expand. The Libyan government launched the GMMRP in 1983 to pursue this proposal. GMMRP aims to transport groundwater from the interior to the coast for agricultural and urban development while reducing extraction from stressed coastal aquifers (Hamad and Saaid 2022; Mcloughlin 1991; Salem 1992; GWA 2006; GMMRA 2022)). According to the main systems and subsystems (connections), as shown in Fig. 8 and Table 2, the plan was to transport approximately 6.5 million cubic meters of freshwater daily via massive buried pipes (prestressed concrete cylinder pipes PCCP, with a diameter of 4 m) to cover the water demands for urban, agricultural, and industrial purposes.

The GMMRA, based in Benghazi, manages the GMMRP, which is Libya's primary water source for urban and agricultural use. Water is delivered to cities via turnouts and tanks linked to municipal distribution networks (GCWW). Currently, agricultural usage makes up the majority of the GMMRP's water use (57.50%), followed by urban use (37.50%) and industrial use (1.40%). It is unclear how quantities are distributed among different users. The amounts of water From GMMRP have increased gradually since the arrival of the 'first water from the first and second phases in 1989 and 1996 (GPC 1999), respectively, reaching 608.56 million cubic meters in 2017 (UNICEF 2019). The number of people supplied with water from the GMMRP currently exceeds about 4 million. The GMMRP has provided water for irrigation and urban purposes since 1989 and 1996, despite inadequate funding for operation and maintenance.

The urban water supply sources from groundwater, as shown in Fig. 9 and Table 3, are divided into; springs, local water wells, and GMMRP. There are about 450 natural springs in Libya, concentrated primarily in the Al Jabal Al-Akhdar, Jabal Nafusa region and southwest Libya. These springs contribute to the minimal urban water supply in sparsely populated areas, mainly suburbs and villages.

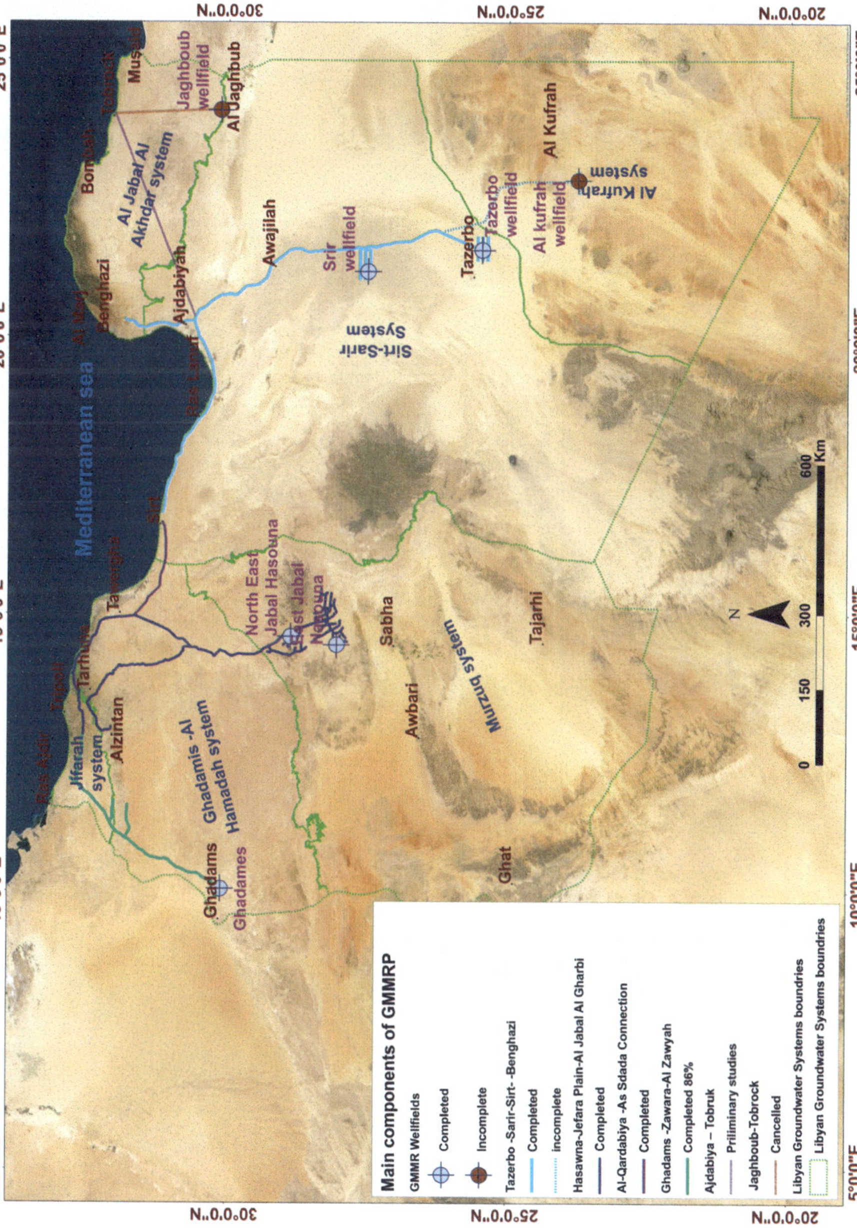

Fig. 8 Map of the main components of GMMRP map based on data from (GMMRA 2022)

Table 2 Main water conveyance systems and connections of GMMRP (GMMRA 2022)

	System	Designed transport volume m^3/day	Urban m^3/day	Agricultural m^3/day	Industrial m^3/day	Completion %
Tazerbo-Sarir-Sirt-Benghazi	Eastern	2,000,000	950,000	1,000,000	50,000	100
Al Kufrah-Tazerbo	Eastern	1,650,000	0	1,500,000	0	70
Ajdabiya-Tobruk	Eastern	Connection	700,000	0	10,000	Preliminary studies
Jaghboub-Tobrock	Eastern	–	–	–	–	Cancelled
Hasouna-Jefara plain	Western	2,500,000	1,200,000	1,250,000	50,000	100
Al-Qardabiya-As Sdada connection	Western	980,000	–	–	–	100
Ghadams-Zawara-Al Zawyah	Western	250,000	250,000	–	–	86

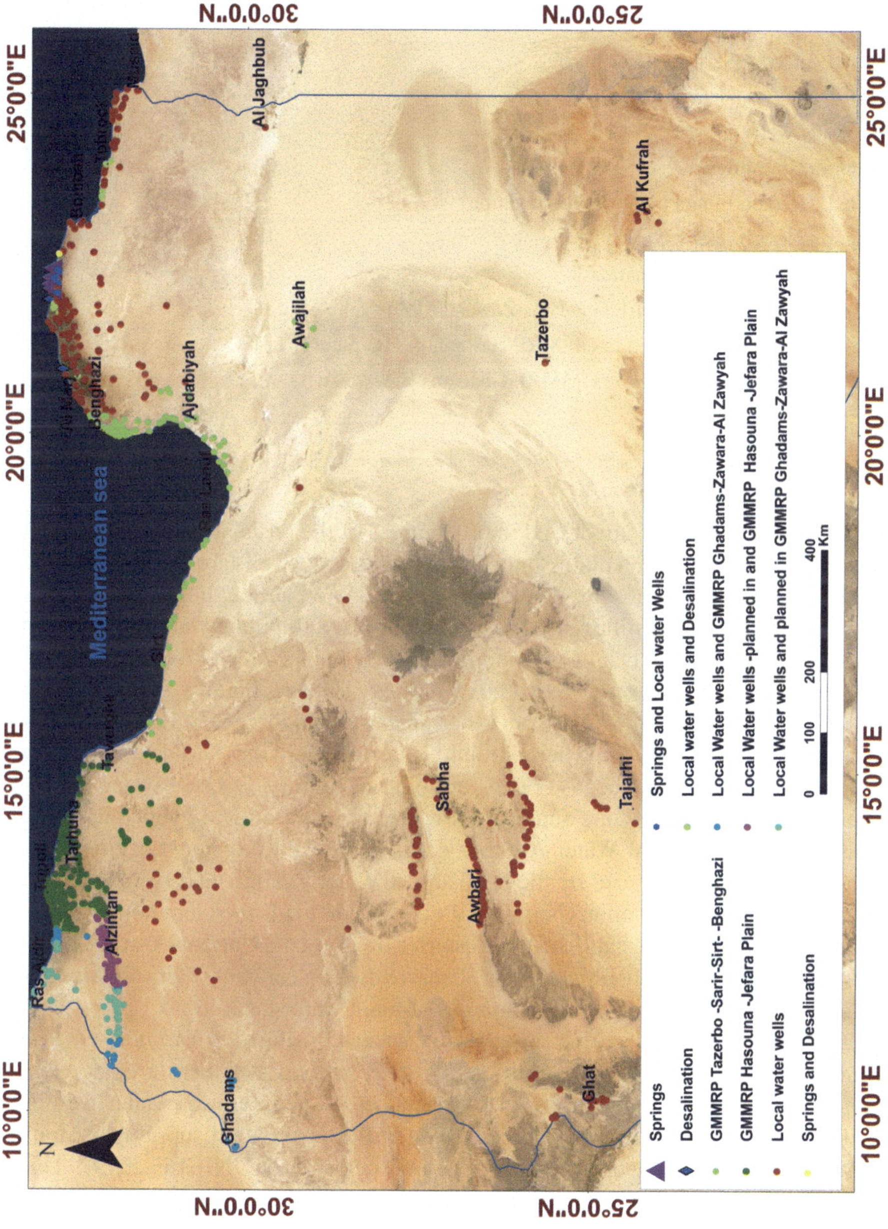

Fig. 9 Map for urban settlements and water supply sources

Table 3 Counts of urban settlements and water supply sources

Urban water supply source	Count of urban settlements
Springs	5
Desalination	30
GMMRP Tazerbo-Sarir-Sirt-Benghazi	84
GMMRP Hasouna-Jefara plain	265
Local water wells	285
Springs and desalination	20
GMMRP Hasouna-Jefara plain and desalination	1
Springs and local water wells	18
Local water wells and desalination	11
Local water wells and GMMRP Ghadams-Zawara-Al Zawyah	28
Local water wells and planned in GMMRP Hasouna-Jefara plain	124
Local water wells and planned in GMMRP Ghadams-Zawara-Al Zawyah	46
Total	917

Most local water wells are drilled in urban areas of the Al Jabal Al Akhdar, Jabal Nafusa areas and the Jifarah plain. Only 53.7% of the 1591 wells are active. Accurate water supply coverage rates are not available in GCWW, although the 2006 census indicates that 572,309 families or 64.5%, are connected to public water distribution networks (BCS 2020). Private wells supply 154,007 households (17.4%) with rain-water and vehicle-borne water to 139,911 homes (15.8%). Finally, the remaining 2.3% of households obtain water from other unspecified sources (UNICEF 2019).

Accurate information on where groundwater through the GMMRP is being distributed is unavailable because of institutional instability. Data on water extraction and consumption became scarce following Libya's political events and civil war in 2011. The General Water Authority was responsible for most of the water information and was replaced by the Ministry of Water Resources (GWA) in 2012. Furthermore, during the country's political division in 2014, the General Authority for Water Resources was established in western Libya and the National Water Corporation in eastern Libya. Obtaining comprehensive data on Libyan water statistics has become challenging with this fragmentation of main water institutions.

Water meters to measure the amounts of water consumed are only used to a minimal extent in some cities. Additionally, there are no statistics on water loss owing to leaks, illegal connections, or other causes. The amount of water lost due to leakage was calculated to be between 30 and 50% of the total supply per capita and the amount of water used inside the home per capita (UNICEF 2019). This large amount of wasted water reflects the poor condition of the water distribution networks. However, the following reports provide information on water demand and consumption:

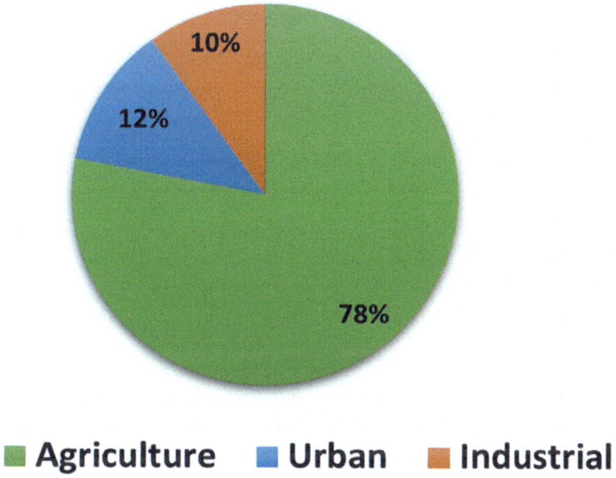

Fig. 10 Water use in Libya (GWA 2006)

- The report on the Great Jamahiriya's water situation and the national strategy for water resource management by the General People's Committee—Technical Committee for the Study of the Water Situation (GPC 1999).
- Water Situation report for the Great Jamahiriya published by the General Water Authority (GWA 2006).
- The National Water Security Strategy report by the National Planning Council which was based on the two previous reports (NPC 2014).

According to GWA (2006), agriculture use in Libya accounts for 78% of total water use, while urban use is 12%, and industrial use is 10%, as shown in Fig. 10. Water consumption is increasing in Libya due to population growth, urbanization, and economic activities. Irrigated agriculture is expanding along the coast, as well as in oases and along wadis.

Urban water consumption varies greatly, ranging from less than 150 L/cap./d in small rural settlements to more than 350 L/cap./d in large cities (NPC 2014; GWA 2006; Pallas 2001). The (GPC 1999) report concluded that the quantities of water consumed for urban use were about 452.56 MCM in 1999, as illustrated in Table 4 and Fig. 11. The urban consumption was 41% in the Jifarah plain and 27% in AI Jabal Al Akhdar, 7% in Sirt-Sarir and AI Kufrah, and 25% in Gadamis-Al Hamadah and Murzuq. The urban consumption was also divided according to water sources: Local water wells were 54%, GMMRP 42%, and seawater desalination was 4%.

In 2006 the water consumed for urban use was about 603 MCM, and the consumption was divided according to water sources as follows: Local water wells at 36%, GMMRP at 53%, and seawater desalination at 11% (GWA 2006). However, because of the development plan for urban water supply (GPC 1999), the contribution of the GMMRP and seawater desalination has increased, and the contribution of local water wells has decreased since 2006. In 2017 according to (UNICEF 2019), the GCWW provided a total of about 575.0 MCM; 59% from local water wells, 59%,

Table 4 Available urban water supply MCM/year in 1998(GPC 1999)

Area	Average urban water supply MCM/Year			Total	Average consumption per capita (liter/person/day)
	GMMRP	Local water wells	Desalination		
AI Jabal Al Akhdar	61.5	52.7	5.1	119.3	254
Sirt-Sarir and AI kufrah	18.6	11.2	0.5	30.3	320
Jifarah plain	109.5	78.6	0	188.1	248
Gadamis-Al Hamadah	0	44.3	12	56.3	180
Murzuq	0	58.06	0	58.06	492
Total	189.6	244.86	17.6	452.06	

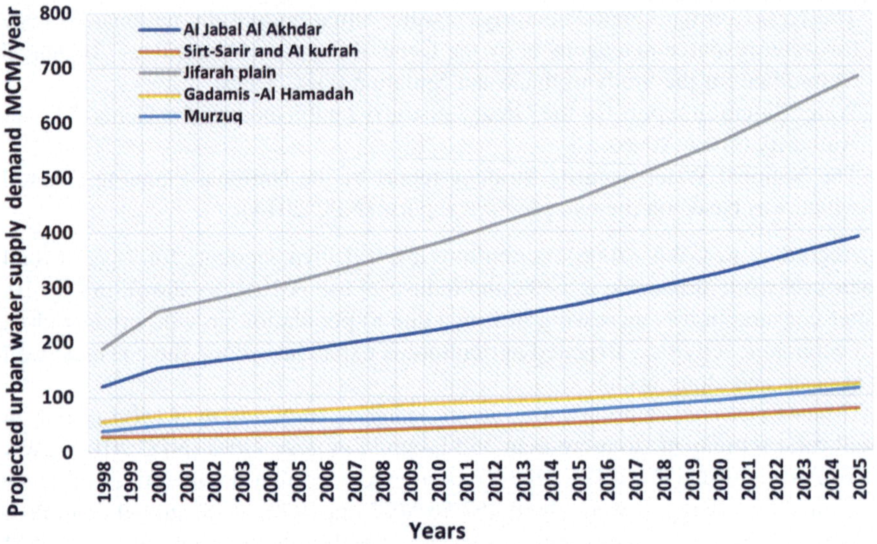

Fig. 11 Projected urban water supply demand 1998 to 2025 (GPC 1999)

32% from GMMRP, and 9% from seawater desalination 52.1 MCM. The GMMRP's diminished supply is discussed in the next section, the urban water supply challenges.

A projection of urban water supply demand from 1998 to 2025 assumes a constant rate of population growth in Libya over the years (1954–1999) and per capita 300 L/cap./d (GPC 1999) (Fig. 11; Table 5). Another projection has an estimated demand of 1200 MCM for 2025 (GWA 2006). Moreover, the third future projection for urban water supply demand from 2015 to 2050 is shown in Fig. 12 and Table 6 (NPC 2014). This projection is based on reducing groundwater supplies for agriculture, Libya's

Table 5 Projected urban water supply demand 1998 to 2025 in MCM/year (GPC 1999)

Year	AI Jabal Al Akhdar	Sirt-Sarir and AI kufrah	Jifarah plain	Gadamis-Al Hamadah	Murzuq	Total
1998	119.3	28.98	188.1	56.8	38.51	431.69
2000	152.85	30.3	255.69	67.91	47.76	554.51
2005	184.18	35.26	311.09	76.46	58.06	665.05
2010	221.94	42.89	378.48	89.09	59.23	791.63
2015	267.44	52.19	460.48	96.93	73.46	950.5
2020	322.26	63.49	560.25	109.13	91.1	1146.23
2025	388.33	77.25	681.63	122.87	112.99	1383.07

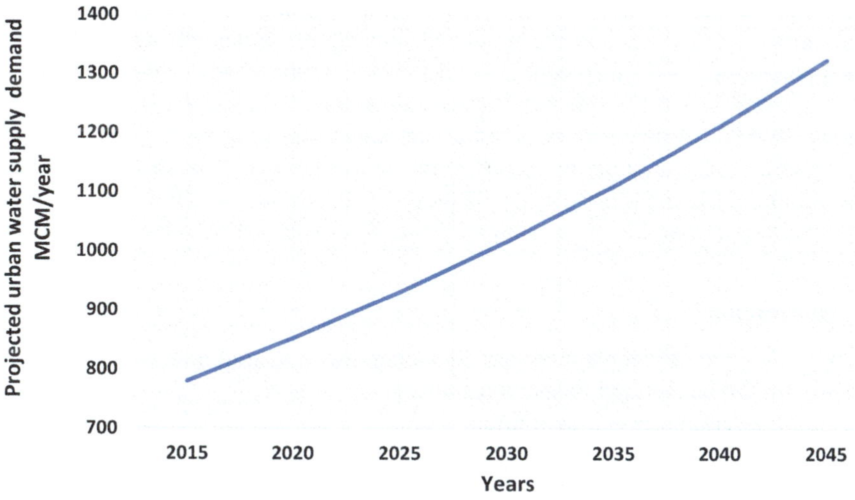

Fig. 12 Projected urban water supply 2015 to 2050 in MCM/year (NPC 2014)

largest water use consumer, to 50%. Unfortunately, these projections were based on underestimations of population and the future increase in urban water supply demand. Also, the quantities from local water wells are inaccurate, as no flow meters are installed on water wells.

5 Urban Water Supply Challenges

Long-standing water shortages in Libya have recently been made worse by population increases and infrastructure failure. Libya now faces a water supply shortage to meet its demands, and the urban and suburban population is rising, whereas in many suburban areas, the water supply is only available every two weeks and only

Table 6 Projected urban water supply demand 2015 to 2050 in MCM/year (NPC 2014)

Years	Projected urban water supply demand MCM/year
2015	714.55
2020	780.45
2025	852.43
2030	931.04
2035	1016.91
2040	1110.69
2045	1213.13
2050	1325.01

once a week in most urban areas. Moreover, climate change is a reality that affects communities worldwide, and the predicted impacts of climate change, including rising temperatures, shifting precipitation patterns, extreme weather, and flash floods, may exacerbate water demands in Libya (Ali and Hafi 2022; Hamed et al. 2018; El-Tantawi 2005). In Libya, the issue of urban water supply is more complex than just a problem of a quantitative shortfall in the availability of fresh water. Urban water supply in Libya faces many challenges and constraints, which we can attempt to outline here (UNICEF 2019; Hamad 2012; Abufayed and Elkebir 2010; GWA 2006):

(1) **Institutional**

- Fragmentation in planning and financing due to institutional instability
- Limited budget and financial resources
- The overlap of water institutions
- The lack of coordination between the water institutions and the lack of integrated management
- Organizational instability
- Limited structuring among the functional levels
- Inadequate institutional capacity at regional and local levels
- Disengaged stakeholders

(2) **Human**

- Socio-economic dimensions are insufficiently reflected
- Insufficient capacity in water management
- Insufficient skilled labour
- Absence of an organized long-term approach to awareness-raising activities
- Lack of financial incentives for workers in water institutions

(3) **Legislative**

- Some laws conflict with water resources conservation

- The adoption of laws governing the protection of some water resources disassociated with implementation procedures and without economic alternatives for affected parties
- The enforcement of water laws made by different parties without integration between parties

(4) **Water sources**

Local water wells

- High transportation costs as the deep aquifers where groundwater exists are far from the population centers
- Groundwater in most urban areas is up to 300 m deep, resulting in prohibitive drilling costs and large-capacity pumps.
- Some areas' agricultural expansion resulted in a decline in groundwater levels and depletion, particularly in the shallow, high-quality aquifers.
- The lack of hydrogeological data on existing water wells from the relevant institutions of the municipalities and the urban settlements
- Some urban water wells' groundwater is unsuitable for human consumption due to its poor quality.
- The absence of routine water quality tests and lack of labs with the necessary equipment in most urban areas.
- Delay in completing the water wells contracted to be implemented on the dates stated.
- Seawater intrusion in coastal aquifers has resulted from haphazard extraction in coastal urban areas.
- The vast majority of water wells are without flow meters, so there is no data for groundwater withdrawals.
- The lack of commitment by some municipalities to the specifications necessary for implementation, supervision, and follow-up by the relevant government water institutions.
- The lack of groundwater protection zones to protect the good quality and quantity aquifers and the lack of law enforcement tools and legislative measures.
- Expansion of random water wells drilling in the urban areas and the lack of sewage infrastructure, thus, the aquifer's risk of pollution. In some areas, pollution has already occurred.
- There are risks of groundwater pollution in most urban areas due to the lack of sewage treatment plants; most of the existing plants are not working, and the existing sanitation infrastructure in large cities is more than 40 years old and in poor condition.
- Lack of qualified and knowledgeable operators to operate water well pumps without instructions or sufficient training.
- In many urban areas, the water well is connected directly to the distribution network without a collection tank, which may cause malfunctions due to the increased pressure on the submersible pump.

GMMRP

- Armed conflict, insecurity, and lawlessness have put GMMRP workers, infrastructure, and facilities at considerable risk. The harm to people includes fatalities, injuries, robberies, and the forced abandonment of construction sites.
- Limited financial resources, including problems acquiring letters of credit and the availability of foreign currency.
- Power outages and fuel shortages.
- Inability to collect revenues from water sold for agricultural and urban purposes.
- Due to the nature, size, and interconnected operation of the infrastructure components of the GMMRP, if one segment fails, all subsequent segments of the system will experience service interruptions.
- The GMMRP infrastructure is extremely remote over vast areas of Libya, making protection difficult.
- Some equipment is designed specifically for GMMRP and is not widely available on the local market; destroyed or damaged components are challenging to replace or acquire.
- Possible corrosion of prestressed concrete pipes due to damage to the cathodic protection systems and theft of its components.
- Technical support from foreign companies and consultancy offices has diminished since 2011 because of poor security.
- Many control chambers, valves, and pipes have suffered damage due to unauthorized connections resulting in significant water theft and loss.
- Some components of the GMMRP deteriorated after 2014 due to the devastating damage caused by armed conflict, insecurity, lawlessness and instability.

6 Conclusion and Recommendations

The need to ensure a comprehensive and reliable water supply for Libya's urban centers is urgent. A consistent potable water supply has always been an essential foundation for economic growth, good public health, and political stability. Libya is predicted to have significant impacts from climate change, including rising temperatures, shifting precipitation patterns, extreme weather, and flash floods. As a result, the resilience of the water supply and sanitation systems will be weakened.

Despite many of the difficulties and issues that were previously discussed, local water wells are a precious source and should be maintained and enhanced. If constructed and developed following the proper criteria, they can contribute up to one-third of the demands of Libya's urban areas. The GMMRP remains the most crucial water supply and is expected to provide 2.3 billion cubic meters annually. However, because of the political climate, it is currently only able to supply 900 million cubic meters annually.

Wastewater reuse for irrigation must be developed, as it can provide a significant amount of freshwater and extend the life of the aquifers from which the GMMRP extracts its water. In that case, the project can satisfy the urban water needs of most of Libya's population for several decades. However, despite its significance as a national strategic project, the GMMRP does not receive adequate support.

Because of the urgent demand for urban water, the Libyan government must reduce the agricultural need. The previous agricultural development plans relied on non-renewable groundwater resources. It is imperative to preserve, develop, and manage water resources sustainably to close the significant gap between the supply and the rising demand due to population growth. Alternative sustainable solutions such as desalination of seawater and wastewater treatment are also needed. Additionally recommended are the following measures:

1. **Develop the conventional and non-conventional water resources:**
    - Update data and information related to hydrogeological units and aquifers.
    - Establish groundwater and surface water monitoring networks.
    - Complete hydrogeological and geophysical studies to inform sustainable groundwater use. Complete implementation of planned dams and water harvesting structures.
    - Protect the springs in Libya and ensure their water resources are used sustainably.
    - Expand the establishment of seawater desalination plants.
    - Complete the implementation of sanitation plans and projects.
2. **Reducing the water deficit:**
    - Impose agricultural regulations to limit the production of crops with high water requirements.
    - Increase the efficiency of irrigation systems.
    - Promote agricultural cycles that require minimal irrigation.
    - Promote rainfed crops.
    - Promote economic activities with low water demand.
    - Raise public awareness of water issues.
3. **Protect water resources from pollution by taking the following measures:**
    - Reduce groundwater depletion, especially in coastal zones.
    - Prepare and apply specifications for the use and disposal of treated water, both urban and industrial.
    - Provide treatment and drainage of water used in urban and industrial areas.
    - Rationalize the use of chemicals for agricultural purposes.
    - Protect surface water estuaries.
    - Support the regulatory authorities concerned with protecting the environment.
    - Use industrial technologies that are more water efficient.
    - Complete environmental impact studies for public and industrial projects.
    - Enact legislation to control the discharge of water associated with oil extraction and ensure compliance.

4. **Developing human and institutional capacities:**
   - Implement continuous capacity building and rehabilitation programs for workers in the water sector.
   - Offer the option for postgraduate studies in hydrogeology and groundwater management to ensure adequate competence in groundwater management.
5. **Ensure that urban purposes, mainly interior areas are given priority access to the GMMRP.**

References

Aboelnga HT, Ribbe L, Frechen FB, Saghir J (2019) Urban water security: definition and assessment framework. Resources 8(4):1–19. https://doi.org/10.3390/resources8040178

Abufayed A, Elkebir A (2010) Water supply and sanitation in Libya: gap analysis, national needs assessment and UNDP interventions

Ahmed M (2020) Sustainable management scenarios for northern Africa's fossil aquifer systems. J Hydrol. https://doi.org/10.1016/j.jhydrol.2020.125196

Aimar A (2017) Managing water crisis in the North African region: with particular reference to Jijel region. In: Behnassi K, McGlade M (ed), Environmental change and human security in Africa and the Middle East, 1st ed. Springer, Cham, p 19. https://doi.org/10.1007/978-3-319-45648-5_12

Aivar C, Francois M (2016) Groundwater governance in the Middle East and North Africa. Groundwater governance in the Arab world, 1st ed. International Water Management Institute (IWMI). https://publications.iwmi.org/pdf/H048385.pdf

Alamin SA (2012) Development of a sustainable water management strategy in the Alwahat area of Libya. Nottingham Trent

Ali AAM, Hafi ZB (2022) Standardized precipitation index (SPI) in north Libya and connection with north Atlantic oscillation (NAO). Quantum J Eng Sci Technol 3(1):14–28. https://qjoest.com/index.php/qjoest/article/view/48

Bakhbakhi M, Salem O (1999) Why the Great Man-made river project? In: General Water Authority & Jamahiriya LA (eds) Regional aquifer systems in arid zones—managing non-renewable resources. UNESCO, pp 3–16

BCS (2020) Estimated Libyan population by regions for the year (2020). Libyan Bureau of Census and Statistics. http://bsc.ly/?P=5&sec_Id=8&dep_Id=7#144

CEDARE (2014) "Libya water sector M&E rapid assessment report". Monitoring & evaluation for water in North Africa (MEWINA) project, Water Resources Management Program. http://web.cedare.org/wp-content/uploads/2005/05/Libya-Water-Sector-Monitoring-and-Evaluation-Rapid-Assessment-Report.pdf

El-Tantawi A (2005) Climate change in Libya and desertification of Jifara Plain using geographical information system and remote sensing techniques [der Johannes Gutenberg-Universität]. https://d-nb.info/97597503X/34

FAO (2009) Atlas of natural resources for agricultural use in Libya (LIB/004 (ed.), 1st ed. Food and Agriculture Organization. https://www.fao.org/publications/card/en/c/afa20dd0-7f4a-4e11-beb9-d9b3b79e0e65/

Fienen MN, Arshad M (2016) The international scale of the groundwater issue. In: Jakeman A, Barreteau AJ, Hunt O, Rinaudo RJ, Ross JD (eds) Integrated groundwater management: concepts, approaches and challenges, 1st ed, pp 1–762. https://doi.org/10.1007/978-3-319-23576-9_2

Frédéric F (2020) Groundwater storage changes in the major North African transboundary aquifer systems during the GRACE era (2003–2016). 12(2669):1–21. https://doi.org/10.3390/w12102669

GMMRA (2022) Great Man-Made river project authority. https://gmra.com.ly/index.php/en

Gouadrzi GH (1970) Geology and mineral resources of Libya-a reconnaissance

Gouadrzi (1980) Structure-Libya. In: Salem M, Busrewil MJ (eds) Geology of Libya: Tripoli, vol III. Academic Press, London, pp 879–892

GPC (1999) Great Jamahiriya's water situation and the national strategy for water resource management

Guendouz A, Moulla AS (2010) The shared resources in the North-Western Sahara aquifer system (Algeria-Tunisia-Libya): the use of environmental isotopes (Algeria part). In: Paris (ed), International conference "transboundary aquifers: challenges and new directions" (ISARM2010) (Issue December, pp 8–11). UNESCO

GWA (2006) Water resources status of Libya. General water Authority

Hallett D, Clark-Lowes D (2016) Petroleum geology of Libya, 2nd ed. Elsevier B.V

Hamad S (2012) Status of groundwater resource of Al Jabal Al Akhdar Region. North East Libya 1(1):11

Hamad S (2020) Updating topographic maps at scale 1:250000 for Libyan territory using quantum GIS (QGIS) and open geospatial data: Libya topo-project. J Geogr Stud 4(1):22–34. https://doi.org/10.21523/gcj5.20040103

Hamad S, Ahweej Y (2020) Evaluation of the Nubian sandstone aquifer system (NSAS) in Al Kufra Oasis, Southeast Libya. Desalination and Water Treatment. https://doi.org/10.5004/dwt.2020.25503

Hamad S, Saaid F (2022) Characterization and management evaluation of the nubian sandstone aquifer in Tazerbo wellfield of the Libyan man-made river project. Appl Water Sci. https://doi.org/10.1007/s13201-022-01684-6

Hamed Y, Hadji R, Redhaounia B, Zighmi K, Bâali F, El Gayar A (2018) Climate impact on surface and groundwater in North Africa: a global synthesis of findings and recommendations. Euro-Mediterr J Environ Integration 3(1):1–15. https://doi.org/10.1007/s41207-018-0067-8

IGRAC U-I (2016) Transboundary aquifers of Africa. Based on: transboundary aquifers of the world [map]. Edition 2015. Scale 1 : 50000000 (2nd ed.). https://www.un-igrac.org/sites/default/files/resources/files/TBAmap_Africa_2016.pdf

Jones JR (1971) Ground water provinces in Libyan Arab Republic. Geology of Libya, Faculty of Science, University of Libya, Tripoli, pp 449–457

Klitzsch E (1971) The structural development of parts of north Africa since Cambrian time. In: Gray C (ed) First symposium on the geology of Libya. Faculty of Science, University of Libya, pp 253–262

MacDonald AM, Bonsor HC, Dochartaigh BÉÓ, Taylor RG (2012) Quantitative maps of groundwater resources in Africa. Environ Res Lett 7(2):7. https://doi.org/10.1088/1748-9326/7/2/024009

MacDonald AM, Lark RM, Taylor RG, Abiye T, Fallas HC, Favreau G, Goni IB, Kebede S, Scanlon B, Sorensen JPR, Tijani M, Upton KA, West C (2021) Mapping groundwater recharge in Africa from ground observations and implications for water security. Environ Res Lett 16(3). https://doi.org/10.1088/1748-9326/abd661

Mcloughlin PFM (1991) Libya's Great Manmade river project prospects and problems. Nat Res Forum 15(3):10–13. https://doi.org/10.1111/j.1477-8947.1991.tb00136.x

NPC (2014) National water security strategy. https://npc.gov.ly

Pallas (1980) Water resources of the socialist people's Libyan Arab Jamahiriya. The geology of Libya. In: Second symposium on the geology of Libya, Held at Tripoli, September (12–16) 1978, pp 539–593

Roudi-Fahimi F, Creel L, de Souza R-M (2000) Finding the balance: population and water scarcity in the Middle East and North Africa. https://u.demog.berkeley.edu/~jrw/Biblio/Eprints/PRB/files/FindingTheBalance_Eng.pdf

Salem O (1992) Libya's future water supply The Great Manmade River Project A partial solution to Libya's future water supply. Int J Water Resour Dev 8:37–41. https://doi.org/10.1080/07900629208722564

Selley RC (1997) The sedimentary basins of Northwest Africa: stratigraphy and sedimentation. In: Selley RC (ed) African basins: sedimentary basins of the world 3. Elsevier, pp 3–16

Shaki A (2006) Simulation and optimization studies for better groundwater management in The Murzuq Basin, Libya (Issue 1) [Heriot-Watt University]. http://www.secheresse.info/spip.php?article9457

Swailem FM, Hamza MS, Aly AIM (1984) Isotopic composition of groundwaters from Kufra (Libya) as indicator for groundwater formation. Isotopenpraxis Isotopes Environ Health Studies 20(2):64–68. https://doi.org/10.1080/10256018408544979

UN (2018) World urbanization prospects. https://population.un.org/wup/publications/Files/WUP2018-Report.pdf

UNECA (2005) Libya national water development report. https://knowledge.uneca.org/SRONA/cmis/views/9fb15c25-614b-4e05-86db-0c74fd7c3dd7%3B1.0

UNICEF (2019) Assessment of water supply systems and institutions in Libya. https://www.humanitarianresponse.info/sites/www.humanitarianresponse.info/files/documents/files/assesment_of_national_water_systems-english.pdf

Wheida E (2012) The water resources management in Libya. Libyan Agric Res Center J Int 3(3):144–154. https://doi.org/10.5829/idosi.larcji.2012.3.3.1209

World Bank (2020) Urban population (% of total population)—Libya. United Nations Population Division. World Urbanization Prospects: 2018 revision. https://data.worldbank.org/indicator/SP.URB.TOTL.IN.ZS?locations=LY

WWAP (2015) The United Nations world water development report 2015: water for a sustainable world. https://www.undp.org/saudi-arabia/publications/un-world-water-development-report-2015?utm_source=EN&utm_medium=GSR&utm_content=US_UNDP_PaidSearch_Brand_English&utm_campaign=CENTRAL&c_src=CENTRAL&c_src2=GSR&gclid=Cj0KCQjwyt-ZBhCNARIsAKH11779aTm84l6Hja7fB8q

Mr. Salah Hamad is an assistant professor of Hydrology and Geoinformation Science at the Department of Natural Resources, Faculty of Natural Resources and Environmental Science, University of Omer Al-Mukhtar, Al Bayda, Libya. He holds an M.Sc. in Geographic Information Systems and Science from Salzburg University, Salzburg, Austria, in 2008. Mr. Hamad has been actively teaching undergraduate courses and supervising bachelor's degree students' graduation since 2010. His research interests lie in GIS mapping and spatial analysis, with a particular focus on geoinformatics research, geodata documentation, and the application of various GIS and remote sensing techniques in conventional water resources management, engineering, and other related aspects of earth science.

Rod Fensham Ph.D., is an associate Professor School of Biological Sciences Faculty of Science-University of Queensland, Australia He is an expert on springs. He has developed a network of other experts, A Fellowship of the Spring, who promote sustainable groundwater use and seek to understand and preserve springs and oases worldwide.

Rainwater Harvesting in Libya: Sustainable Solutions for Water Scarcity

Hamdi A. Zurqani, Mhi Eldin M. Alkhboli, Salah Hamed, Abdulsalam Al-Bukhari, and Abdulazim S. Fadil

Abstract Rainwater harvesting presents a viable solution to address water scarcity challenges in arid regions such as Libya. This chapter explores the potential of rainwater harvesting in Libya, a country characterized by limited water resources due to low precipitation levels and reliance on finite water sources. Traditional methods such as rooftop collection and modern techniques, including dam construction offer opportunities to augment water resources sustainably. However, ensuring the safety and reliability of harvested rainwater is essential, necessitating comprehensive water quality testing and treatment protocols. Despite challenges such as low rainfall and high evaporation rates, innovative solutions based on modern technologies and sustainable practices can enhance water resilience. Selected case studies illustrate practical implementation and outcomes, providing valuable insights for plans. Recommendations include allocating resources for infrastructure development, educating communities, establishing water quality protocols, prioritizing dam construction, strengthening capacity, and formulating supportive policies. Through the implementation of these measures, Libya can effectively address water scarcity challenges and ensure water security for present and future generations.

H. A. Zurqani (✉)
University of Arkansas Agricultural Experiment Station, Arkansas Forest Resources Center, College of Forestry, Agriculture, and Natural Resources, University of Arkansas, Monticello, AR, USA
e-mail: Hzurqani@uark.edu; Zurqani@uamont.edu

M. E. M. Alkhboli
Department of Soil and Water Sciences, Faculty of Agriculture, University of Omar Al-Mukhtar, Al Bayda 00218, Libya

S. Hamed · A. Al-Bukhari
Department of Forest and Rangeland, Faculty of Natural Recourses and Environmental Sciences, University of Omar Al-Mukhtar, Al Bayda 00218, Libya
e-mail: a.albukhari@omu.edu.ly

A. S. Fadil
Agriculture Research Center of Libya, Al Bayda 00218, Libya

H. A. Zurqani (ed.), *Water Resources of Libya*, Springer Water,
https://doi.org/10.1007/978-3-031-80920-0_4

Keywords Water scarcity · Arid regions · Sustainable practices · Water resilience · Infrastructure development · Policy frameworks · Innovative technologies

1 Introduction

Rainwater harvesting is an alternative water source worldwide in arid and semi-arid regions (Jackson et al. 2001; Lancaster 2019). Yet, the amount of rainfall in these areas is limited and highly variable, resulting in low water productivity (Molden 2013). Moreover, this scarce water is further exacerbated by increased human population growth, urbanization and a growing diversification of water users (World Bank 2023). Therefore, there is a dire need to introduce rainwater harvesting techniques that can significantly augment the quantity of rainwater being gathered and stored as opposed to the current limitation (United States Environmental Protection Agency 2012). The limitation can be overcome by installing large-capacity storage-tanks or underground cisterns to accommodate the maximum gathering of rainwater to ensure its availability for domestic use, irrigation, or other purposes (Datta 2019). In addition to these techniques, soil contouring—leveling—or slope terracing provides an alternative measure in increasing rainwater infiltration rates into the soil (Ngigi 2003) and, in turn, increasing water productivity since groundwater levels are replenished. Hence, rainwater harvesting can best be maximized with such measures in place. Consequently, these regions can become self-sufficient and not dependent on scarce water resources. Therefore, investments in such systems will be an essential part of a sustainable water management approach and conservation of natural resources in the country.

Libya is almost entirely covered by the Libyan Desert, a flat plateau that forms a part of the Sahara (Zurqani 2021). As a result of its desert climate, the country receives very minimal rainfall with only a narrow strip along the coastline receiving more than 100 mm of rain annually (Ben Mahmoud and Zurqani 2021). This low rainfall makes water a scarce resource in the country. As a result, Libya has had to rely heavily on underground water sources and desalination plants to meet its water needs. However, the constant extraction of groundwater has led to concerns over the depletion of this vital resource. Rainwater harvesting is a traditional water conservation method used in Libya for centuries. It involves collecting rainwater from rooftops, courtyards, and other surfaces and storing it for later use (UNICEF 2022). Rainwater harvesting can also be used for various purposes, including drinking water, irrigation, and livestock watering. However, despite its potential benefits, rainwater harvesting alone is not enough to address the water scarcity issues in Libya. The construction of major dams in Libya has provided another essential solution to water scarcity. These major dams in Libya have a total capacity of 385 mm^3 and an average annual storage capacity of 61 mm^3 (Brika, 2019). These dams play an important role in harvesting rainwater and storing it for various uses such as agricultural water supply, industrial projects, and domestic use.

Without proper analysis and monitoring of harvested rainwater, there is a risk of potential threats and hazards associated with its use (Igbinosa and Aighewi 2017). One of the potential threats is the presence of harmful pollutants or contaminants in the harvested rainwater, which can pose health risks to humans and animals if consumed. These contaminants include bacteria, viruses, heavy metals, pesticides, and chemicals (U.S. Environmental Protection Agency 2000; Taylor and Barrett 2013). Thus, there is a need to implement effective water quality testing and treatment methods to ensure the safety and suitability of the harvested rainwater for its intended purposes. Regular monitoring should be conducted to detect any contaminants or pollutants that may have entered the water during collection or storage (Capehart et al. 2021). By addressing these risks and maintaining the quality of harvested rainwater, the dams in Libya can continue to be a reliable and sustainable solution to water scarcity in the region. This can be achieved by implementing comprehensive water quality testing protocols encompassing chemical and microbiological analysis (International Organization for Standardization ISO Central Secretariat 2020). Moreover, using advanced filtration systems and disinfection methods, such as ultraviolet (UV) treatment, can further enhance the safety of the harvested rainwater. By regularly inspecting and maintaining the dams and storage facilities, any potential sources of contamination can be identified and addressed promptly, ensuring the continued suitability of the water for agricultural, domestic, and industrial purposes (World Health Organization 2022).

The problem lies in the lack of knowledge and resources among farmers and communities to implement suitable techniques for water harvesting. In addition, some water harvesting methods require testing under current environmental conditions (Vohland and Barry 2009; Nachshon et al. 2016; Singh and Turkiya 2017). Moreover, the socioeconomic and cultural factors in different regions further complicate the adoption of water harvesting techniques. Therefore, it is crucial to educate farmers about these techniques and provide them with the necessary resources and support to successfully implement water harvesting practices (Krejci and Beamon 2014). Thus, enhancing the capabilities of both communities and national research programs and extension services is needed in the water harvesting area. Furthermore, promoting the adoption of water harvesting techniques in agricultural practices is crucial for agricultural development, especially in arid and semi-arid areas (Danquah et al. 2019). Water harvesting techniques have been proven effective in increasing productive yields without depleting groundwater sources in African countries (Krejci and Beamon 2014). The successful implementation of water harvesting techniques in similar regions has encouraged the adoption of these approaches on a large scale in Libya.

There are many different types of rainwater harvesting systems that can be used in Libya. Some popular options include rooftop harvesting systems, where rainwater is collected from rooftops and stored in tanks for later use (Singh and Turkiya 2017). Another common system is the surface runoff water harvesting system, which collects rainwater from paved surfaces and stores it in reservoirs or underground tanks (Nachshon et al. 2016). Wadi rainwater harvesting is also an effective method of accumulating precipitation using dry riverbeds known as wadi (Noureddine et al. 2021).

This process involves impeding water flow via check dams or other constructions, effectively permitting water collection (Noureddine et al. 2021). The collected water can be stored in tanks or cisterns or used for irrigation, facilitating crop cultivation within the Wadi. Groundwater recharge systems also replenish underground water sources by diverting rainwater into the ground through wells or infiltration trenches (Kebede and Taye 2021). These various rainwater harvesting systems offer practical solutions for addressing water scarcity and promoting sustainable water management in Libya.

2 Selected Case Studies in Rainwater Harvesting in Libya

2.1 Selection and Characterization of Integrated Benchmark Research Watersheds in Libya

In 2011, Ziadat and Oweis conducted a research study on behalf of the International Center for Agricultural Research in the Dry Areas (ICARDA) to develop and implement a process for the selection and characterization of integrated benchmark research watersheds (IBRWs) in Libya. The project determined study areas in eastern and western Libya based on agricultural region mapping results and workshop discussions. In this study, the data was collected, verified for accuracy, reclassified based on selection criteria, and analyzed to match biophysical and socioeconomic conditions of watersheds in study areas. The analysis included various factors such as rainfall, cropping systems, rural settlements, accessibility and visibility, topography, and soil. As a result of this analysis, 16 watersheds in the east and 18 watersheds in the west were identified as potentially suitable for achieving the project objectives. An interdisciplinary team of researchers visited all the watersheds identified and discussed their potential during field visits and meetings. Eventually, they identified seven watersheds (3 in the west and 4 in the east) as the most suitable for the project activities (Table 1).

Four watersheds have been chosen to characterize and implement project activities (watersheds 83 and 416 in the western area and watersheds 17 and 37 in the eastern

Table 1 The numbers and local names by region for the seven selected watersheds (adapted from: Ziadat and Oweis 2011)

No.	Watersheds numbers and local names by region	
	Al-Jabal Al-Gharbi (Western region)	Al-Jabal Al-Akhdar (Eastern region)
1	Watershed no. 83 (Al-Ghadama)	Watershed no. 37 (Samalos)
2	Watershed no. 99 (Saffeat)	Watershed no. 58 (Al Qatara)
3	Watershed no. 416 (Al-Nakaza)	Watershed no. 28 (Al Mualaq)
4	–	Western part of watershed no.17 (Al Marj)

area) Fig. 1a and b. The primary objective of the project is twofold. Firstly, in the short term, it aims to help identify suitable locations for rainwater harvesting, supplemental irrigation interventions, and other management practices. Secondly, in the long term, it aims to establish a database that can be used by scientists not only during the project but also after its completion. Suitability criteria include both biophysical and socioeconomic parameters. Field surveys were conducted to gather missing information, followed by synthesizing maps and layers of information. The final step was to combine all the information into a format that would meet the needs of the project (Ziadat and Oweis 2011).

The results of the watershed characterization process have revealed that integrating multiple biophysical elements, such as watershed characteristics, land suitability, and hydrological characteristics, with socioeconomic characterization, such as community distribution and characteristics, accessibility, and willingness to cooperate, is essential to develop and achieve research watershed objectives (Ziadat and Oweis 2011). It is also worth pointing out that this approach is highly reproducible and can be used in other areas. Overall, this study provides a valuable approach for selecting watersheds in Libya for integrated research on rainwater gathering and supplementary irrigation. The selected watersheds can offer a unique opportunity to test and evaluate various technologies under different conditions. Furthermore, the outcome of this study is expected to contribute to improving agricultural productivity and water management in Libya.

2.2 *Opportunities and Challenges for Groundwater Artificial Recharge by Surface Water Harvesting in Libya*

Hamad (2012) conducted a study on the opportunities and challenges for groundwater artificial recharge surface water harvesting in Libya. The study discussed the c opportunities and challenges of surface water harvesting in Libya during investment and development. The study showed several water harvesting projects have occurred over the past decades in Libya. Owing to limited water resources, Libya was divided into seven regions to maximize utilization of surface water and seasonal wadi water (Table 2).

The study has highlighted the importance of establishing dams in the group of wadis that empty their water into the sea. The study suggests that 18 major dams should be implemented to protect towns, villages, and agricultural and industrial businesses from flood hazards (Hamad 2012). These dams will also help collect water for agricultural and domestic purposes, prevent soil drift, and create water bodies for agricultural use. Aquifer feeding has been introduced in Libya's most important wadis, with a total storage capacity of 389 million cubic meters and an annual average storage volume of about 6 million cubic meters of water (General Water Authority 2009) (Table 3).

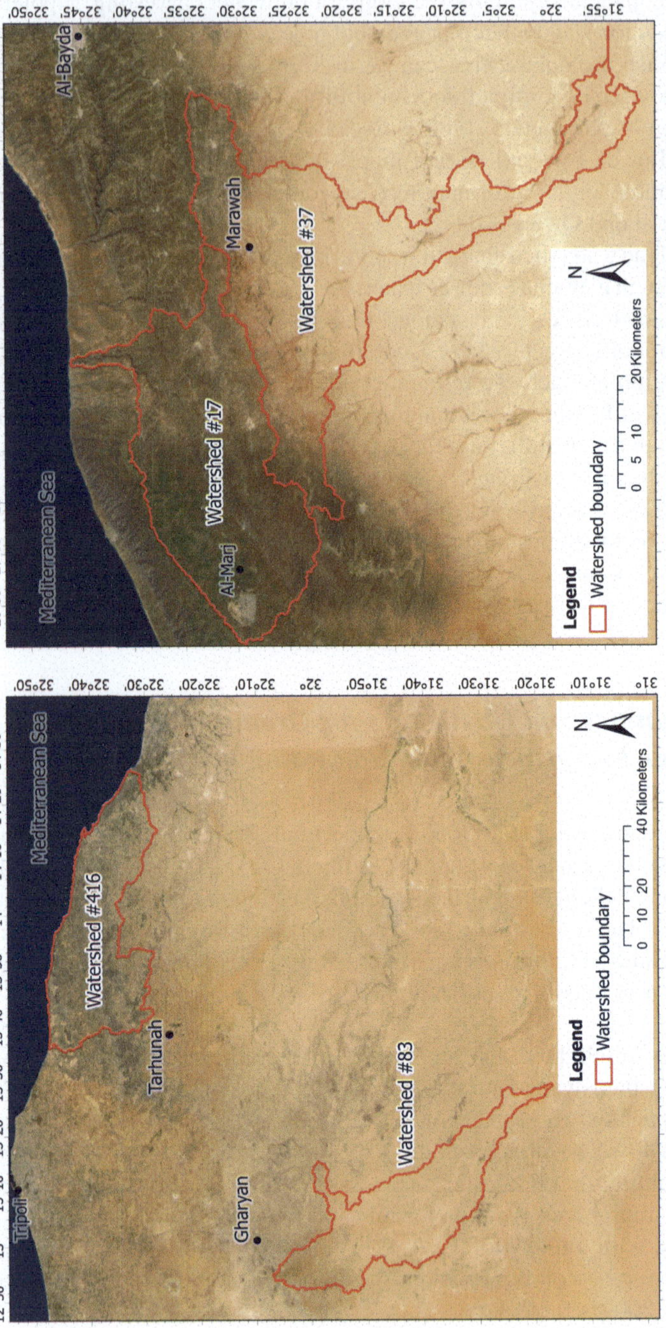

(a) The selected watersheds in the western area (watersheds 83 and 416)

(b) The selected watersheds in the eastern area (watersheds 17 and 37)

Fig. 1 The selected watersheds for this study: **a** the three selected watersheds in the western area, and **b** the four selected watersheds in the eastern area

Table 2 Surface water quantities (million cubic meters) that can be reserved and expected to be reserved in the future (General Water Authority 2009)

Region	Total storage capacity	Controlled in 1977	Controlled in 2009	Expected future control
Jafara plain	87	10.10	25.50	47.30
Musrat Al Khomas	27	13.50	17.40	20.76
Ghdames Basin	16	–	–	13.10
Central Libya	3	0.90	1.80	2.10
Al Jabal Al Akhdar	45	13.60	15.95	35.90
South Al Jabal Al Akhdar	11	–	–	–
Tobrock	11	–	–	1.35
Total	200	38.10	60.65	120.51

Table 3 Permeable dams (General Water Authority 2009)

No.	Wadi name	Region	Reservoir capacity mm^3	Average annual design storage mm^3 year
1	Wadi Mejenin	Bin Ghasherr	58	10
2	Wadi Kaam	Zliten	111	13
3	Wadi Ghan	Giryan	30	11
4	Wadi Zaret	Rabtah	8.6	4.5
5	Wadi Lebda	AlKhoumes	5.2	3.4
6	Wadi Qattara	Benghazi	120	12
7	Wadi Qattara ‖	Benghazi	1.5	0.50
8	Murqus	Ras Al Hilal	0.15	0.15
9	Bin Jawad	Bin Jawad	0.34	0.34
10	Zaza	Tukrah	2	0.80
11	Derna	Derna	1.15	1
12	Abu Mansur	Derna	23.7	2
13	Wadi Tabrit	Zliten	1.6	0.50
14	Wadi Al Thiker	Zliten	2.4	0.50
15	Wadi Jarif	Sirt	2.4	0.30
16	Wadi Zabid	Sirt	2.6	0.50
17	Wadi Zahawuiyah	Sirt	2.2	0.70
18	Wadi Shouka	Jufrah	3.65	0.20
Total			376.49	61.39

Table 4 Dams contracted for implementation (General Water Authority 2009)

No.	Wadi name	Region	Reservoir capacity mm^3	Average annual design storage mm^3 year
1	Wadi Bu Shyba	Giryan	5.5	2.58
2	Wadi Ar Ruman	Giryan	3.75	1.4
3	Wadi Tlal	Sirt	8.7	–
4	Wadi Al Hammar	Benghazi	19.5	5.85
Total			37.45	9.83

Table 5 Dams ready for contracting (General Authority for Water)

No.	Wadi name	Region	Reservoir capacity mm^3	Average annual design storage mm^3 year
1	Wadi Al Khalij	Derna	6	2.5
2	Wadi Al Mualaq	Derna	5	3
3	Wadi Ashahbuon	Trahuna	3.3	1.02
4	Wadi Az Zgadnah	Trahuna	1.9	0.62
5	Wadi Turgut	Qarabouli	8.4	2.08
6	Wadi Qraim	Qarabouli	2.6	0.62
7	Wadi Ghanima	Qarabouli	2.4	0.63
Total			29.6	10.47

The study also indicated that the survey of other significant wadis on which dams can be erected has contracted the implementation of 4 major dams, as shown in Table 4. Technical studies and specifications of seven dams have been prepared in different regions of Libya, as shown in Table 5. It was proposed that 13 more dams be implemented after the completion of their study, as in Table 6.

The study identified the primary challenges and limitations faced by surface water harvesting projects in Libya, which can be summarized as follows:

- Centralization of planning and financing and lack of budgets and necessary financial resources.
- Overlap between water institutions, lack of coordination and integration, instability of their organizational structure, and lack of capacity development programs in surface water management.
- Lack of influential stakeholders' involvement, social and economic dimensions not considered in planning and implementation, and lack of a long-term approach to outreach activities.

Table 6 Proposed dams (General Authority for Water 2009)

No.	Wadi Name	Region	Reservoir capacity mm^3	Average annual design storage mm^3 year
1	Wadi Bani Walid	Bani Walid	10.40	7.20
2	Wad Atmisalah	Bani Walid	9.3	3.25
3	Wadi Mansour	Bani Walid	4.25	1.80
4	Wadi Mimon	Bani Walid	3.40	0.85
5	Wadi Nalut	Nalut	5.90	1.25
6	Wadi Bu Ar Rasef	Al Hrabah	15.00	2.28
7	Wadi um Al Qrab	Ar Hybat	10.00	1.55
8	Wadi Jenawen	Jadou	5.00	0.89
9	Wadi Tobrock	Tobrock	2.35	1.35
10	Wadi Swakh	Giryan	6.00	2.60
11	Wadi Bu Aysha	Giryan	2.80	1.30
12	Wadi Al Bab	Benghazi	31.30	8.80
13	Wadi Sirt-Bin Jwad	Sirt	1.25	1.25
Total			106.88	33.65

Overall, the study concluded that the surface water resources in Libya are limited to the country's northern regions and are small quantities that can be developed and invested by implementing proposed dam projects and other water harvesting techniques. Artificial groundwater feeding can be done under a pilot program, including the data control and collection period.

3 The Benefits of Rainwater Harvesting in Libya

Rainwater harvesting is an effective way to provide a sustainable water source in Libya. This is particularly important in rural areas with limited access to freshwater sources. This practice can help to reduce the pressure on existing water supplies and decrease the dependency on imported water (Vohland and Barry 2009). Moreover, this practice promotes self-sufficiency and resilience in water scarcity, making communities less vulnerable to drought and water shortages (Mutekwa and Kusangaya 2007). Furthermore, rainwater harvesting is an environmentally friendly practice that helps to conserve natural resources and reduce the strain on traditional water sources (Biazin et al. 2012). Overall, there are many benefits to rainwater harvesting in Libya, including:

- **Increased water availability**: Rainwater harvesting can help increase water availability in Libya. This is especially important in areas where water is scarce.

- **Reduced reliance on other water sources**: Rainwater harvesting can help to reduce dependence on other water sources, such as groundwater and surface water. This can help to protect these water sources from overuse.
- **Improved water quality**: Rainwater is generally of good quality and does not require treatment before it can be used. This can save money on water treatment costs.
- **Environmental benefits**: Rainwater harvesting can help to reduce erosion and improve soil moisture. It can also help to recharge groundwater supplies.

4 The Challenges of Rainwater Harvesting in Libya

Rainwater harvesting in Libya is challenging due to the country's arid climate and limited rainfall. The irregular and unpredictable rainfall in the region makes it difficult to plan and implement effective rainwater harvesting systems (Ammar 2017). Furthermore, the lack of proper infrastructure and technology further hampers the efficient collection and storage of rainwater (Wehrey et al. 2023). In addition, the quality of harvested rainwater is a significant concern, as dust and other contaminants in the atmosphere can affect its cleanliness, making it less suitable for consumption and other purposes (Deng 2021). These challenges to rainwater harvesting in Libya can be summarized as the following:

- **Low rainfall**: Libya has a low average annual rainfall, making rainwater harvesting difficult.
- **High evaporation rates**: Libya has high evaporation rates, which can also make rainwater harvesting difficult.
- **Water quality**: The quality of rainwater can be affected by pollution, so collecting rainwater from a clean source is important.
- **Soil salinity**: The soil in Libya is often saline, making it difficult to grow crops with rainwater.
- **Maintenance**: Rainwater harvesting systems require regular maintenance to ensure that they are functioning properly.
- **High cost**: The cost of installing a rainwater harvesting system can be high, especially for large systems.

Innovative solutions and a comprehensive approach to rainwater harvesting in Libya are necessary to address these challenges. Modern technology and sustainable practices can be leveraged to overcome these obstacles and utilize rainwater as a valuable resource in the country.

5 The Future of Rainwater Harvesting in Libya

Despite the challenges, rainwater harvesting is a promising water resource for many people in Libya. With increasing water scarcity, rainwater harvesting is becoming an increasingly important way to meet the water needs of households, businesses, and communities. It can help to increase water security, reduce demand for groundwater, improve water quality, reduce flooding, and create jobs. With more awareness and support, rainwater harvesting can play a significant role in meeting Libya's future water needs. Several things can be done to promote rainwater harvesting in Libya, including:

- **Providing financial assistance for rainwater harvesting systems**: This can help to make rainwater harvesting more affordable for households and businesses.
- **Providing technical assistance for rainwater harvesting systems**: This can help to ensure that rainwater harvesting systems are installed and maintained properly.
- **Educating the public about the benefits of rainwater harvesting**: This can help to raise awareness of rainwater harvesting and encourage its use.

6 Conclusion and Recommendations

Rainwater harvesting offers a vital solution to the enduring water scarcity challenges faced by arid regions like Libya. With its limited rainfall and reliance on nonrenewable groundwater resources, Libya exemplifies the urgent need for innovative water management strategies. Traditional techniques such as rooftop collection and dam construction, alongside modern methods like groundwater recharge systems, present opportunities to manage water resources sustainably.

However, ensuring the safety and reliability of harvested rainwater remains paramount. Comprehensive water quality testing and treatment protocols, coupled with robust monitoring and maintenance, are essential to safeguard public health and sustain water resources. Despite challenges such as low rainfall, high evaporation rates, and soil salinity, innovative solutions leveraging modern technologies and sustainable practices offer pathways to resilience.

Selected case studies in Libya illustrate the practical implementation and valuable outcomes of rainwater harvesting initiatives, providing valuable insights for future endeavors. The multifaceted benefits of rainwater harvesting, including enhanced water availability, reduced reliance on finite sources, and environmental sustainability, underscore its transformative potential in shaping Libya's water future.

Through concerted efforts, strategic investments, and widespread adoption of rainwater harvesting techniques, Libya can navigate the challenges of water scarcity and forge a sustainable water future for generations to come. By promoting awareness, providing financial and technical support, and fostering community engagement, rainwater harvesting can emerge as a cornerstone of Libya's water resilience strategy, ensuring a reliable and equitable water supply. Several recommendations emerge to enhance its effectiveness and sustainability:

1. Allocate resources for developing rainwater harvesting infrastructure, including large-capacity storage tanks, underground cisterns, and surface runoff collection systems. This investment will facilitate the maximum gathering and storage of rainwater, ensuring its availability for various purposes.
2. Educate farmers and communities about suitable rainwater harvesting techniques and provide necessary resources and support for their implementation. Focus on promoting techniques such as rooftop harvesting, surface runoff collection, and wadi rainwater harvesting, tailored to the socioeconomic and cultural context of different regions.
3. Establish comprehensive water quality testing and treatment protocols to ensure the safety and suitability of harvested rainwater. Regular monitoring of rainwater harvesting infrastructure is essential to detect and address any contaminants or pollutants, safeguarding public health and sustainability.
4. Emphasize integrated watershed management approaches for rainwater harvesting projects. Utilize interdisciplinary teams to select and characterize suitable watersheds, considering biophysical and socioeconomic factors and optimizing the effectiveness of rainwater harvesting initiatives.
5. Prioritize the construction of major dams in strategic locations to harvest rainwater and store it for various uses, including agricultural supply, industrial projects, and domestic consumption. Implementing dams in wadis and dry riverbeds can help prevent flood hazards while creating water reservoirs for sustainable water management.
6. Strengthen the capacity of national research programs, extension services, and local communities in rainwater harvesting techniques and management. Foster collaboration among stakeholders, including government agencies, research institutions, NGOs, and local communities, to enhance the effectiveness and sustainability of rainwater harvesting initiatives.
7. Formulate supportive policies and regulations to promote rainwater harvesting and allocate adequate funding for its implementation. Provide financial incentives, subsidies, and technical assistance to encourage the adoption of rainwater harvesting systems, particularly in rural and water-stressed areas.

References

Ammar AA (2017) Evaluating rainwater harvesting systems in arid and semi-arid regions. Doctoral dissertation, Wageningen University and Research. https://research.wur.nl/en/publications/evaluating-rainwater-harvesting-systems-in-arid-and-semi-arid-reg

Ben Mahmoud KR, Zurqani HA (2021) Soil forming factors and processes. In: The soils of Libya, pp 33–48

Biazin B, Sterk G, Temesgen M, Abdulkedir A, Stroosnijder L (2012) Rainwater harvesting and management in rainfed agricultural systems in sub-Saharan Africa–a review. Phys Chem Earth, Parts A/B/C 47:139–151

Brika B (2019) The water crisis in Libya: causes, consequences and potential solutions. Desalination and Water Treatment

Capehart MA, Artiola JF, Eden S (2021) Preparing rainwater for potable use. Retrieved from https://extension.arizona.edu/pubs/az1863-2021.pdf
Danquah FO, Twumasi MA, Asare I (2019) Factors influencing Zai pit technology adaptation: the case of smallholder farmers in the upper east region of Ghana. Agric Res Technol 21:13–24
Datta PS (2019) Water harvesting for groundwater management: issues, perspectives, scope, and challenges. John Wiley & Sons
Deng Y (2021) Pollution in rainwater harvesting: a challenge for sustainability and resilience of urban agriculture. J Hazardous Mater Lett 2:100037
General Water Authority Libya–GWA (2009) Report of achievements. Unpublished Report
Hamad SM (2012) Opportunities and challenges for groundwater artificial recharge by surface water harvesting in Libya. Libyan Agric Res Centre J Int 3(6):260–268
Igbinosa IH, Aighewi IT (2017) Assessment of the physicochemical and heavy metal qualities of rooftop harvested rainwater in a rural community. Global Chall 1(6):1700011
International Organization for Standardization ISO Central Secretariat (2020) Water quality—Sampling—Part 15: Guidance on the preservation and handling of sludge and sediment samples (ISO 5667-15:2009). https://www.iso.org/standard/42892.html
Jackson RB, Carpenter SR, Dahm CN, McKnight DM, Naiman RJ, Postel SL, Running SW (2001) Water in a changing world. Ecol Appl 11(4):1027–1045
Kebede S, Taye MT (2021) Groundwater scarcity and management in the arid areas in East Africa. In: Global groundwater. Elsevier, pp 177–186
Krejci C, Beamon B (2014) Environmentally-conscious supply chain design in support of food security. Oper Supply Chain Manage: Int J 3(1):14–29
Lancaster B (2019) Rainwater harvesting for drylands and beyond, volume 1: guiding principles to welcome rain into your life and landscape (vol 1). Rainsource Press
Molden D (ed) (2013) Water for food water for life: a comprehensive assessment of water management in agriculture. Routledge
Mutekwa VT, Kusangaya S (2007) Contribution of rainwater harvesting technologies to rural livelihoods in Zimbabwe: the case of Ngundu ward in Chivi District. Water SA 32(3). https://doi.org/10.4314/wsa.v32i3.5270
Nachshon U, Netzer L, Livshitz Y (2016) Land cover properties and rain water harvesting in urban environments. Sustain Cities Soc 27:398–406
Ngigi SN (2003) Rainwater harvesting for improved food security. http://repository.seku.ac.ke/handle/123456789/3168
Noureddine G, Eslamian S, Katlane R (2021) Integrated water resources management and climate change. J Int Sci et Technique de L'eau et de L'environnement 1(1):89–103
Singh O, Turkiya S (2017) Assessing potential for rooftop rainwater harvesting: an option for sustainable rural domestic water supply in arid region of Haryana. J Rural Dev:49–60
Taylor S, Barrett M (2013) Long-term performance and life-cycle costs of stormwater best management practices. In: Novatech 2013–8ème Conférence internationale sur les techniques et stratégies durables pour la gestion des eaux urbaines par temps de pluie/8th International Conference on planning and technologies for sustainable management of Water in the City. GRAIE, Lyon, France
The United Nations Children's Fund (UNICEF) (2022) UNICEF Libya water scarcity and climate change: an analysis on WASH enabling environment in Libya. https://www.unicef.org/mena/documents/unicef-libya-water-scarcity-and-climate-change-analysis-wash-enabling-environment-libya
U.S. Environmental Protection Agency (2000) Fact sheet: methodology for deriving ambient water quality criteria for the protection of human health. https://www.epa.gov/wqc/fact-sheet-methodology-deriving-ambient-water-quality-criteria-protection-human-health-revised
United States Environmental Protection Agency (2012) Guidelines for water reuse. EPA 600-R-12–618. USEPA. https://www.epa.gov/waterreuse/guidelines-water-reuse
Vohland K, Barry B (2009) A review of in situ rainwater harvesting (RWH) practices modifying landscape functions in African drylands. Agr Ecosyst Environ 131(3–4):119–127

Wehrey F, Dargin J, Mehdi Z, Muasher M, Yahya M, Kayssi I, Hassan Z, Andrews M, Madain M, Al-Mailam M, Hamzawy A (2023) Climate change and vulnerability in the Middle East
World Bank (2023) Water. World Bank, Washington, DC. https://www.worldbank.org/en/topic/water/overview
World Health Organization (2022) Guidelines for drinking-water quality: incorporating the first and second addenda. World Health Organization. https://www.ncbi.nlm.nih.gov/books/NBK579458/
Ziadat F, Oweis T (2011) Selection and characterization of integrated benchmark research watersheds in Libya. ICARDA, Aleppo, Syria, xi + 67 pp
Zurqani HA (ed) (2021) The soils of Libya. Springer

Dr. Hamdi A. Zurqani is an Assistant Professor of Geospatial Science in Natural Resource Management and Conservation at the University of Arkansas Agricultural Experiment Station, Arkansas Forest Resources Center, University of Arkansas at Monticello, USA. With a strong academic foundation and a passion for innovative technologies, he is also an FAA (i.e., Pt107) Certified sUAS/Drones Pilot, leveraging this expertise to enhance his proficiency in remote sensing, GIS, and geospatial data analysis, which he applies to advance his research and teaching in environmental monitoring and analysis. Dr. Zurqani is a recognized expert as a result of his internationally acclaimed work in the areas of environmental information science, remote sensing, geospatial analysis, land evaluation, sustainability, pedology, and soil science education. He has conducted research across the world, including the United States of America and Africa, and has served as PI, co-PI, or co-investigator on several grants-funded research projects. In addition to his research and teaching responsibilities, Dr. Zurqani is actively involved in institutional service at the University of Arkansas at Monticello. He serves on the UAM Campus-wide Artificial Intelligence (AI) Review Committee, the UAM Faculty Assembly Student Engagement Committee, and the UAM Library Committee.

Professionally, Dr. Zurqani contributes to the broader academic community through a variety of leadership and collaborative roles. He is an active member of the Steering Committee of the Institute of Electrical and Electronics Engineers (IEEE) Mediterranean and Middle-East Geoscience and Remote Sensing Symposium (M2GARSS) and serves as an Ambassador for the Geoscience and Remote Sensing Society (GRSS) Chapter of the IEEE in Libya, facilitating communication, collaboration, and knowledge-sharing within the geospatial and remote sensing communities. In recognition of his expertise and substantial contributions to the field, Dr. Zurqani was invited in 2024 to serve on the Technical Program Committee for the IEEE Asia-Pacific Conference on Geoscience, Electronics, and Remote Sensing Technology (AGERS). In 2025, he also joined the IEEE GRSS Technical Committee on Remote Sensing Environment, Analysis, and Climate Technologies (REACT), a platform for scientists and engineers to exchange ideas and knowledge on environmental issues and the impacts of climate change, emphasizing support for sustainable development goals (SDGs) through remote sensing technologies. Beyond these roles, Dr. Zurqani holds the position of UAM Co-Principal Investigator at the Southeast Climate Adaptation Science Center, where he is involved in research initiatives focused on climate adaptation and resilience in the southeastern United States and its surrounding regions.

Dr. Zurqani's editorial and peer-review activities further highlight his commitment to advancing scientific research. He serves on the Editorial Boards of several journals, including *Frontiers in Earth Science* (Geoinformatics), *Discover Soil* (Springer Nature), *Frontiers in Remote Sensing*, and *Remote Sensing* (MDPI). He has also been appointed as a Guest Editor for special issues in several journals, focusing on topics such as advancing spatial prediction of soil properties using remotely sensed data and geospatial artificial intelligence, and UAV agricultural management. Dr. Zurqani's dedication to his field is further demonstrated by his role as a peer reviewer for numerous publishers, including Springer Nature, Nature, Frontiers, Elsevier, and MDPI, among others. He is also the editor of several influential publications that address critical environmental challenges. His notable works include: *The Soils of Libya* (2021), part of Springer's World Soils

Book Series; *Environmental Applications of Remote Sensing and GIS in Libya* (2022); and *Water Resources of Libya: Challenges and Management* (2025), published as part of the Springer Water Series (SPWA). These publications highlight his multidisciplinary expertise and commitment to addressing top environmental issues in Libya and beyond.

Dr. Mhi Eldin M. Alkhboli Ph.D., is a Full Professor at the Department of Soil and Water Sciences at the Faculty of Agriculture, University of Omar Al-Mukhtar, Al Bayada, Libya. He completed his M.Sc. in Soil Physics in 1996 from the University of Omar Al-Mukhtar, Libya, and earned his Ph.D. in Environmental Studies in 2003 from the University of Alexandria. Dr. Alkhboli's primary research and teaching focus includes land use management, land evaluation, remote sensing, and geographical information systems. He has worked at various departments, including the Department of Soil and Water, Faculty of Agriculture, Faculty of Natural Resources, Department of Geography, Faculty of Art, University of Omar Al-Mukhtar (2003–2013), Department of Science, Environmental Engineering and Department of Urban and Regional Planning, Graduate Academy Benghazi Branch (2010–2023). Dr. Alkhboli provided technical assistance in establishing the GIS Unit at the Agricultural Research Center in Tubruk (2007) and Al Bayada (2010). He also worked as part of the Water harvesting team for the joint project between ICARDA and Agricultural Research Center, Libya (2010–2012). Dr. Alkhboli has several scientific research papers published in peer-reviewed scientific journals and conferences and has supervised numerous master's theses in the field of interest. Additionally, Dr. Alkhboli was commissioned by the Ministry of Higher Education and Scientific Research to establish a Date Palm Tree Research Center and serves as its general director (2013–2024).

Mr. Salah Hamed is an assistant professor of Hydrology and Geoinformation Science at the Department of Natural Resources, Faculty of Natural Resources and Environmental Science, University of Omer Al-Mukhtar, Al Bayda, Libya. He holds an M.Sc. in Geographic Information Systems and Science from Salzburg University, Salzburg, Austria, in 2008. Mr Hamad has been actively teaching undergraduate courses and supervising bachelor's degree students' graduation since 2010. His research interests lie in GIS mapping and spatial analysis, with a particular focus on geoinformatics research, geodata documentation, and the application of various GIS and remote sensing techniques in conventional water resources management, engineering, and other related aspects of earth science.

Dr. Abdulsalam Al-Bukhari Ph.D., is an Assistant Professor of the Natural Resources Management, Remote Sensing, and Geographic Information Systems at the Department of Forestry and Rangeland, Faculty of Natural Resources and Environmental Sciences, Omar Al-Mukhtar University, Al Bayada, Libya. He received his M.Sc. in Environmental Sciences (rangeland ecology and management) in 2010 from Omar Al-Mukhtar University, Al Bayada, Libya. Dr. Albukhari completed his Ph.D. in Data Risk and Environmental Analytical Methods (The Use of Remote Sensing and GIS for Monitoring Rangeland Degradation in Libya) from Cranfield University, Cranfield, UK, 2019. He specialized in rangeland ecology and management. His main research interests focused on monitoring and assessing rangeland degradation and the environmental information sciences (Remote Sensing and GIS). He is head of the Geographic Information Systems Unit at Omar Al-Mukhtar University. Dr. Albukhari is a member of the Editorial Board for Al-Mukhtar Journal of Sciences.

Mr. Abdulazim S. Fadil is a researcher at the Agricultural Research Center, Al Bayda, Libya. He specializes in developing water harvesting techniques and was the team leader for water harvesting in the agreement between the International Center for Agricultural Research in Dry Areas (ICARDA) and the Agricultural Research Center in Libya. He obtained a Master's degree in Environmental Sciences in 2012 from Omar Al-Mukhtar University, Al Bayda, Libya, specializing in studying water pollution in the Green Mountain region.

Agricultural Drainage in Libya: History and Perspective

Mohamed A. Hamilla and Murad Ellafi

Abstract Drainage in irrigated lands has been installed to manage salinity and waterlogging problems. Less than 10% of all irrigated area requiring drainage is currently drained in developing countries. This chapter aims to (1) review the agricultural drainage development in Libya for the last 70 years. (2) analyze the existing drainage systems and highlights their pros and cons. (3) demonstrate the suggested solutions to improve the drainage systems in Libya. During the 1950s, ditches were randomly dug in fields, and farms suffering from salinity and waterlogging tended to be left. Between 1960–1984, several agricultural projects were established. Still, drainage was neglected or installed without maintenance programs. Therefore, hundreds of hectares were lost due to waterlogging and salinity. In 1985, the General Administration of Irrigation and Drainage (GAID) was established. Between 1985–2017, GAID managed to prepare technical reports based on several field visits and investigations for the current agricultural projects in Libya. The visits covered agricultural projects in three geographical locations: east, central, and southwest Libya. The GAID suggested that installing and maintaining proper drainage systems would help to improve the efficiency of more than 30,000 ha in those geographical locations.

Keywords Agricultural drainage · Irrigated lands · Soil salinity · Waterlogging · Drainage systems · Developing countries · Drainage development

M. A. Hamilla
Irrigation and Drainage Specialist, Tripoli 00218, Libya
e-mail: hamila1960@yahoo.com

M. Ellafi (✉)
Department of Soil, Water, and Climate, University of Minnesota, Twin Cities, MN, USA
e-mail: mellafi@umn.edu

Northwest Research and Outreach Center, University of Minnesota, Crookston, MN, USA

H. A. Zurqani (ed.), *Water Resources of Libya*, Springer Water,
https://doi.org/10.1007/978-3-031-80920-0_5

1 Introduction

Agricultural drainage refers to removing excess water from the land surface and root zone to sustain and improve crop yield. Surface drainage (ditches) and subsurface drainage (drains) become required for filtering excess water in soils where natural drainage is inadequate. Drainage systems have been installed in arid and semi-arid areas to overcome waterlogging and salinity issues. Installing and maintaining drainage systems helps to improve the productivity of many arid and semi-arid lands, such as in the USA (Gassman et al. 2005; Kisekka et al. 2017), Australia (Bennett et al. 2013; Wahba and Christen 2006), India (Gupta et al. 1993; Hirekhan et al. 2007), Turkey (Kale 2012, 2011), Egypt (Kandil et al. 1995; Wahba et al. 2002), China (Chen et al. 2021; Gao et al. 2017) and South Africa (Taguta 2017). However, less than 15% of all irrigated land that requires drainage systems is currently drained, and the proportion of agricultural land drained in developing countries is 5–10% compared to 25–30% in developed countries [Smedema et al. (2000) as cited by Ellafi et al. (2021)]. This is due to low levels of agricultural and rural development, adoption of crops and varieties that can tolerate adverse drainage conditions such as salinity and waterlogging. In addition, the cost of installing and maintaining drainage systems and more traditional (i.e., less innovative, more risk-averse and less competitive) farming attitudes in developing countries compared to developed countries prohibit drainage system installation. In Libya, it was estimated that in 2000, only 9000 ha (around 2% of all croplands) of irrigated land was equipped with some form of drainage (FAO 2016). This is due to a fundamental lack of expertise in the country; lack of knowledge on which drainage models are applicable to arid and semi-arid regions, lack of data for key input parameters for drainage models and the high cost of drainage installation (GWA 2005, 2018).

Since the 1940s, Libya has witnessed a steady increase in agricultural expansion and by the 1950s, water needs in arid Libya had exceeded water resources for food self-sufficiency (Qadir et al. 2003). Also, a large amount of water was lost due to the low efficiency of the irrigation systems and the failure to conduct maintenance operations, and the Corrosion of irrigation equipment (without reaching their lifespan), which led to waterlogging and soil salinity problems in many agricultural lands (Al-Barouni 2013). The absence of a properly maintained drainage system, the high evapotranspiration rates, the low rainfall, and the near-surface depth of the impermeable layer and water table all added to the previous causes. It worsened the waterlogging and salinity situation, converting the land unsuitable for agricultural activities. The need for agricultural drainage in Libya has been realized by stakeholders and decision-makers since the 1950s. However, agricultural drainage was neglected for a while, and subsurface drainage networks were established without regular maintenance programs when it was convinced of its importance. Through extension, reach out to farmers and educate them on the importance of agricultural drainage in arid and semi-arid areas. This chapter aims to (1) review the agricultural drainage development in Libya for the last 70 years. (2) analyses the existing drainage systems in Libya by indicating the type of drainage system installed and their pros and cons. (3) demonstrate the suggested solutions to improve the drainage systems in Libya.

2 Agricultural Drainage Development in Libya

2.1 Agricultural Drainage in the 1950s

During the 1950s, no drainage systems were installed in the country (apart from a few attempts to dig ditches within the agricultural lands). Farms were heavily impacted by waterlogging and salinity (resulting in a spread of Giant reed and Scirpus plants) tended to be left. In addition, cracks on buildings and roads were noticed due to the water table rise. In 1957, the Libyan Ministry of Agriculture started paying attention to the situation by creating pilot fields for more than 40 ha in Tawergha, Brak Eshkeda, and Jufra (GWA 2001, 2018). Experimental studies were conducted in these fields to evaluate the applicability of growing more tolerant crops to salinity. In addition, a team of researchers from Taiwan (National Formosa University) continued the experimental work by installing drainage ditches in agricultural projects in Tawergha, Brak Eshkeda, and Jufra (Hamilla 2018).

2.2 Agricultural Drainage from 1960–1984

Many agricultural projects were developed during this period, but the drainage was forgotten or installed without proper maintenance. As a result, the waterlogging and soil salinity issues increased. Also, the efficiency of the installed drainage systems was low, causing sizeable agricultural lands to be considered not suitable for profitable yield (Hamilla 2005). Due to not having an independent entity for irrigation and drainage in Libya, it was necessary to rely on international experiences in the design and installation of any drainage system within the agricultural projects in Libya. In 1979, A consultancy team from Netherlands was hired to design and install a drainage system in brak Eshkeda and to visit and evaluate the situation et al. Hammam and Al-Furjan agricultural projects in the Jufra district (Al-Khoja and Hamilla 2007). The team assessed the current condition of the projects in Jufra and provided a report with recommendations related to the drainage systems. The team recommended a detailed study and soil survey be conducted on both agricultural projects. Other consultancy teams were hired to design and evaluate the drainage systems within the agricultural lands in Libya (Cornelius-Brochier 1981; Danenco 1980; Holzmann-Wakuti 1974; Italconsult 1976).

2.3 Agricultural Drainage from 1985–2022

A firm conviction was established that irrigation and drainage are essential for Libya's sustainable agricultural development. This resulted from the experience of the agricultural projects implemented between 1970–1984 and the worsening of irrigation and drainage problems. This prompted the officials of the agriculture and water sector to think seriously about forming management concerned with this vital aspect, especially given the absence of an enacting law at the time. In 1985, the General Administration of Irrigation and Drainage (GAID) was established. GAID is responsible for many irrigation and drainage systems within the agricultural projects in Libya (GWA 2001). GAID employs modern techniques to manage water use efficiency for irrigation and drainage schemes in Libya (known as integrated water management for irrigated agriculture). Also, work on repairing, maintaining, and rehabilitating irrigation and drainage projects in Libya and standardizing irrigation and drainage equipment and devices and their accessories under local climate, water, soil, and plant conditions. In addition, developing technical cadres in agricultural water management (especially in irrigation and drainage), and working on using GIS programs in the field to increase the efficiency of irrigation and drainage on regional levels (GWA 2005). In 1991, a Libyan-Dutch group (including engineers from GAID) and FAO members visited and evaluated many agricultural projects. The projects were in different parts of Libya, including the Sabha region (Ghodwa, Samno, and Sabha), Ghat (Awenat, and Tahala), Murzuq (Murzuq, Um-Araneb, Hamera, and Zwela), Jufra (Al-Hammam and Al-Furjan) and Brak Eshkeda. A detailed report was written to highlight the current situation regarding irrigation and drainage, analyses the problems in these projects, and presents a summary with recommendations. The report stated that the priority on improving and installing new drainage systems should be in Al-Hammam, Al-Furjan, Um-Araneb, and Awenat. In 2002, due to the salt accumulation and the water table rise in Al-Hammam, a national company was hired to clean the ditches and remove all the vegetation blocking the water from moving freely (van Aart 1991).

Since its establishment in 1985, GAID has highlighted the problems and implications facing agricultural projects and their irrigation and drainage systems. Many technical reports were written after monitoring the irrigation and drainage systems within the agricultural projects. The reports evaluated the existing agricultural projects in Libya and how the productivity of these projects can be improved by installing and maintaining proper drainage systems. Other technical reports focused on the tools and methods to improve water use efficiency (GWA 2001, 2003, 2005). The main problems in many Libyan agricultural projects are summarized as follows:

- Inappropriate drainage designs.
- Salt accumulation in the soil.
- Irrigation water quality.
- Shallow water table depth.
- Impermeable layer located near the soil surface (1.5 m).

3 Agricultural Drainage and Reclamation of Saline Lands in Libya

3.1 Agricultural Drainage and Reclamation of Saline Lands in the Central Area

3.1.1 Ain Tawergha Agricultural Project

The project is located 35 km south of Misrata city. Figure 1 illustrates the geographical location of Ain Tawergha and the satellite image for the project. Since 1917, locals of Ain Tawergha (Ain means spring water) have used the spring water to irrigate small fields. In 1957, Ain Tawergha was noticed as a project with hundreds of hectares are used for agriculture (Hamilla 2018). Between 1960–1961, the Libyan government requested a soil survey on the project and the survey was done by a team from Durham University. The survey was conducted to evaluate the feasibility of expanding the project (GWA 2018). In 1965, the work continued to assess the possibilities of expanding the agricultural lands and this time by Wakuti (a German consulting company). Wakuti prepared all the required studies and the recommendation designs for the project. In 1970, The General Organization for Agricultural Reclamation and Land Reconstruction (GOARLR) began work on the project's second phase, which includes operations to leach the salt accumulated in the soil (especially in the east part of the project) (GWA 2001). Between 1982–1984, research was conducted to evaluate the feasibility of establishing a dairy-beef project. In 1985, the project started in the western part (more than 51% of the total project area). Alfalfa, barley, and oats were among the crops growing to feed the cattle. However, the project was discontinued in 2001 due to the absence of regular maintenance for the pumping stations, the irrigation system, and the drainage system (GWA 2005). Reclamation was started but never finished, and the project needed more than 3000 ha rehabilitation. The project has open ditches, two main ditches with drain depths up to 3 m, two sub-main ditches with drain depths up to 2.1 m, and field ditches with drain depths up to 1.2 m (Hamilla 2005).

3.1.2 Al-Hammam Agricultural Project

Al-Hammam agricultural project is located on the north-east side of Sokna Oasis (29°04′01″N 15°47′05″E) at an altitude of 270 m ASL (see Fig. 2). Al-Hammam encompasses 1200 ha and is split into 182 farms (6–8 ha each farm). The work on the project began in 1971, and the reclamation was completed in 1985 (GWA 2003). Al-Hammam project has a drainage system designed in the 1970s. The project's drainage system is open ditches dug to remove the excess water from the root zone and leach the salt. The main drainage ditch is located on the east side of the project (Al-Khoja and Hamilla 2007). The main ditch is connected to six sub-drains collecting the

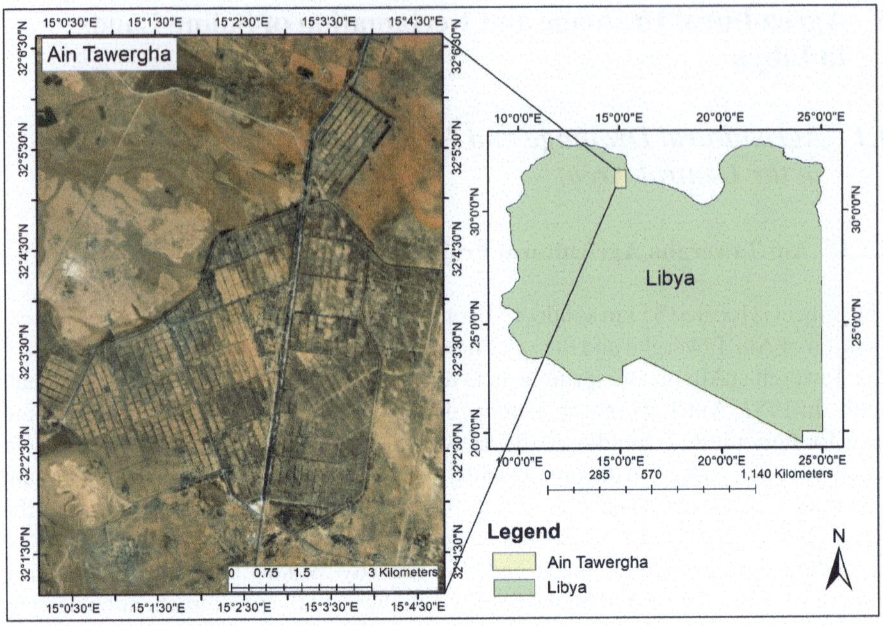

Fig. 1 Satellite image of Ain Tawergha and its location on the map of Libya

water from the field drains and discharging it from west to east. The main ditch discharges the drainage water from the sub-drains to the evaporated pond located north of the project. The area of the evaporated pond is around 33,000 m^2, and the depth is up to 3 m. the total length of the main ditch is about 7.7 km. The total length of the sub drains is around 23 km. currently, the project is suffering from salinity and waterlogging problems (van Alphen and Kor 1984). There is an urgent need for a new drainage system to overcome the issues and improve the yield (Ellafi et al. 2022; Muftah et al. 2006).

3.1.3 Al-Furjan Agricultural Project

The project is located west of Al-Hammam agricultural project and on the north-west side of Sokna Oasis. Al-Furjan covers 2600 ha divided into 256 farms (10 ha for each farm), and around 40 ha was considered a model farm (Fig. 3) (Al-Khoja and Hamilla 2007). The project was designed and implemented by Holzmann-Wakuti in the 1970s (Holzmann-Wakuti 1974). Even though a drainage system is crucial in the project's area, Al-Furjan does not have a proper drainage system installed. Therefore, the water table in the project is shallow, causing soil salinity and waterlogging problems on many farms (GWA 2003, 2007, 2018).

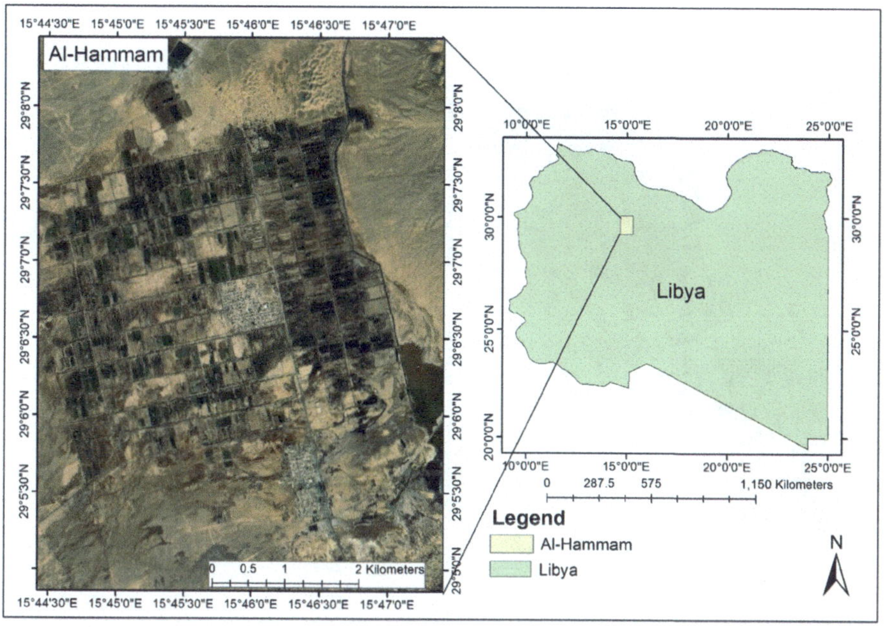

Fig. 2 Satellite image of Al-Hammam agricultural project and its location on the map of Libya

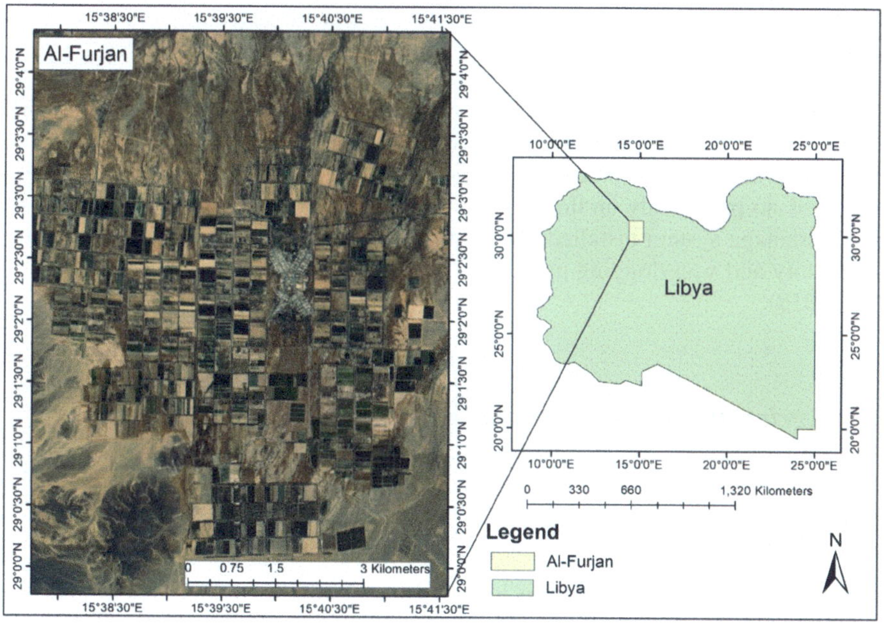

Fig. 3 Satellite image of the Al-Furjan project and its location on the map of Libya

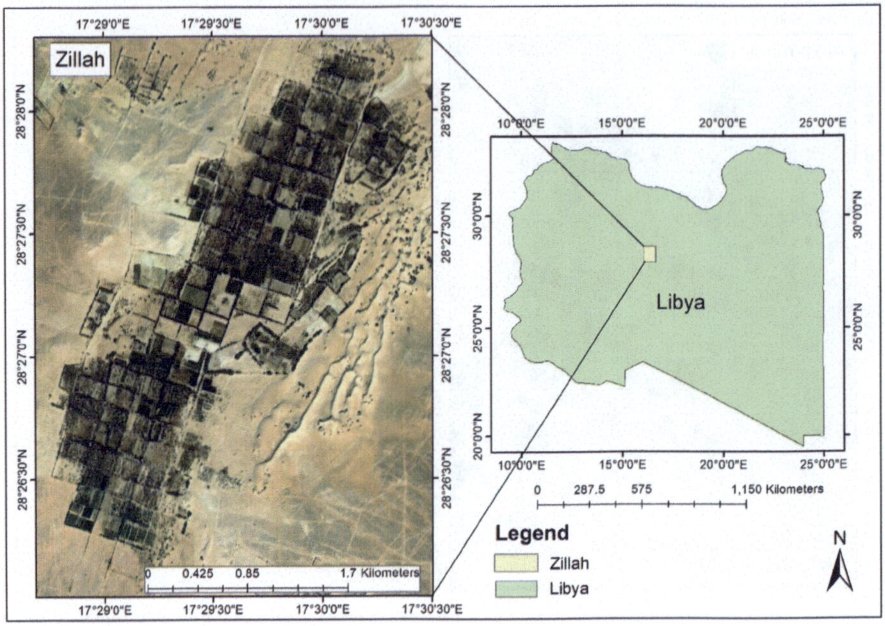

Fig. 4 Satellite image of Zillah agricultural project and its location on the map of Libya

3.1.4 Zillah Agricultural Project

The project is located in Zillah Oasis, south-east Weddan city (Fig. 4). The work began on the project by self-effort from the locals in 1974. The project covers 482 ha, split into 82 farms (5.5–6 ha each). The project mainly grew palm trees and Alfalfa. There was no prior study on the soil and water in the project. The project does not have a drainage system installed; therefore, the project had many issues, including soil salinity and waterlogging problems (Al-Khoja and Hamilla 2007; GWA 2001, 2005, 2018).

3.2 Agricultural Drainage and Reclamation of Saline Lands in the South-West Area

3.2.1 Brak Eshkeda Agricultural Project

The project area of Brak Eshkeda is located 60 km north of Sabha city. The area has about 3000 ha, divided into three sub-areas (see Fig. 5):

- Western Area; 2000 ha
- Central Area; 600 ha

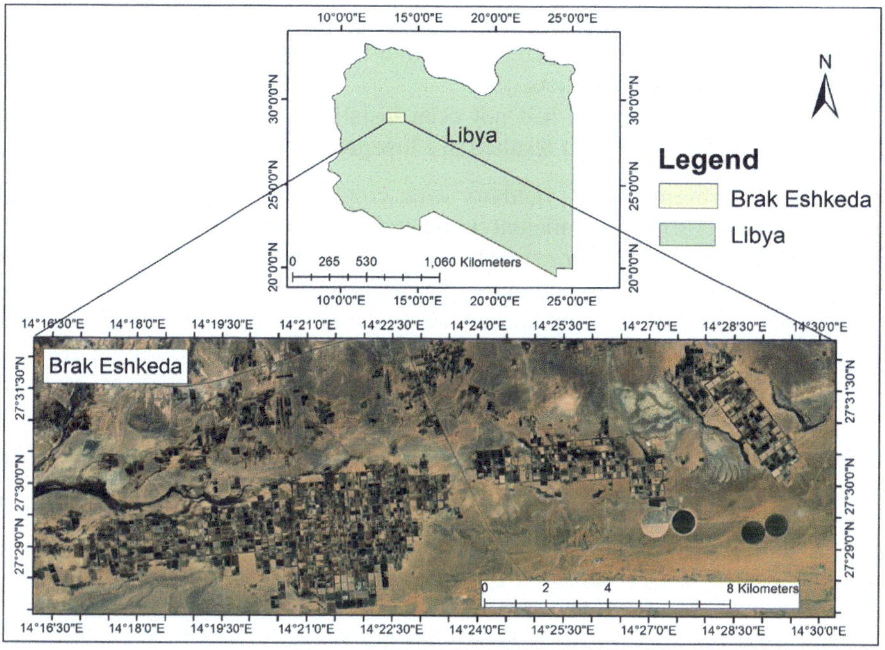

Fig. 5 Satellite image of Brak Eshkeda and its location on the map of Libya

- Eastern Area; 370 ha.

The project implementation started in 1975 and has been under operation since the second half of 1976. In accordance with the installed irrigation wells, the area comprises 25 districts, each with 12 farms (10 ha each). The farms are divided into two plots, each of which is 180 m × 234 m. Windbreaks have been planted between the farm plots and around the farm. Irrigation water is lifted from a deep aquifer (350–400 m depth from the surface). The water is pumped into reservoirs from where it is pumped into the irrigation networks to each farm connected with four hydrants. In 1977, the first drainage problems became visible. In 1981, about 35% of the project suffered from shallow water table depth occurring at less than 1 m depth (Hamilla 2005, 2018).

Due to the raised water table, a drainage system was designed and executed in conformity with the farm size. The general map of the drainage system (prepared by Italiana Lavori in June 1979) shows the layout of the drainage system. The conductor drains (main drains) are open ditches that convey the surplus water to natural depressions (sabkhas) and the valley, where it percolates and evaporates without practical use. The previous drainage design (up to 1983) did not function properly due to the following:

1. Maintenance of ditches is problematic; because of the unstable sandy soils and proliferation of reeds, the open ditches gradually became wider and shallower, resulting in insufficient outlets.
2. Field and collector drains had not been installed, which caused salinity and waterlogging problems and resulted in a low project yield.

In order to prevent loss of yields or even crop failure, the groundwater table has to be drawn down by an efficient drainage system. Drainage aims to remove a surplus of water so that soil structure and aeration are maintained or recovered and to control water tables that would otherwise rise to the near soil surface. Therefore, the government started thinking about finding a solution, including designing and installing subsurface drainage in the project. Cornelius-Brochier was hired in 1981 to design and install subsurface drainage in Brak Eshkeda, which was completed in 1983 (Cornelius-Brochier 1981).

In a subsurface drainage system, one can distinguish three categories of drains (field drains, collector drains and main drains). Field drains serve primarily to control the fluctuation of the groundwater table but may also collect surface runoff. The water in the field drains flows into collector drains which convey the water to the main drainage system, which is carried to the outlet of the area (Hamilla 2018). The main drainage system must provide a free and reliable outlet for the field drains. This is why the previous drainage system did not function properly; therefore, a subsurface drainage system has been proposed and installed (GWA 2001).

The final design for the main drain and field drainage system was carried out in four phases. The area requiring drainage was less than the total area because about one-third is higher levelling land or land with a high hydraulic soil conductivity and a deep groundwater level or not to be drained due to a shallow impervious layer (Hamilla 2005). Table 1 illustrates the four phases that have been taken to design the drainage system in Brak Eshkeda and the drainable area for each phase.

The design parameter and design criteria were:

- The main drains were dimensioned independently according to the gross irrigation area if the field drainage system was executed at a rate of 2.5 mm d^{-1} (This value covers the summer peak drainage rate at a 20% safety margin).
- The hydraulic calculation of the main drain pipes was based on a discharge of 75% of the total flow capacity.
- The minimum gradient applied in the design is 0.5%.

Table 1 The phases for installing a drainage system in Brak Eshkeda

Phase	Total area (ha)	Drainable area (ha)
A	1080	900
B	940	480
C	580	520
D	370	240
A–D	2970	2140

- The minimum pipe diameter was (diameter nominal (DN)) 160 mm, and the maximum diameter was DN 600 mm, fixed in the contract for execution.
- Usually, smooth pipes with a minimum diameter of DN 200, respectively DN 250 mm, must be laid in an open trench. If the drain depth is less than 2.0 m, corrugated PVC pipes with a diameter of DN 160 mm up to 200 mm can be installed by a trenchless drainage machine.
- Control shafts were provided with a spacing of about 250 m on each collector drain outfall and at locations of changing alignments.
- The depth of the main drain depended on the diameter, gradient and depth of the feeder/collector pipe system, particularly at the upstream end of the mains.
- Some drain outlets have been installed to serve maintenance and control drain discharge.

Finally, this is a summary of the layout of the field drainage system installed in Brak Eshkeda:

- The spacing between field drains was adapted to the crops' planting system, primarily orchards planted in a grid system of about 6 m. Thus, the drain spacing was a multiple of 6 m, and the plot size of 234 × 180 m is of a reasonable distance too.
- Field drains were installed lengthwise. Then field collectors were installed broadwise and lengthwise, respectively. According to DIN 1185, the maximum slope is given to the collector drain.
- To connect the field drains to the field collectors, 45° junctions were executed.
- The inspection shafts were constructed at the drain outlet of each farm plot and each angle in the field collector.
- The inspection shafts were designed with a sand trap and had to be hurried at a depth of about 0.5 m below the ground surface.
- The norm discharge of field drainage was set up at 3.0 mm d^{-1}. Only some farm plots surrounded by elevated irrigated land were calculated with an increased discharge of 4.5 mm d^{-1}.
- The optimum drain depth was 1.5 m. Depending on topography, the extreme values range from 0.8 m to 2.0 m.
- The minimum gradient of the field drains and field collectors were 1.0%.
- The feeder pipe diameter was 50 mm (internal diameter 43.9 mm).
- The pipe diameter of collector drains was DN 100, 125, 160, or 200, corresponding to the results of the hydraulic calculation.
- At the lower end of some farm plots, only a perforated collector DN 160 was executed.

3.2.2 Tahala Agricultural Project

Tahala agricultural project covers 1650 ha split into 250 farms (see Fig. 6). The project is located in Tanezzuft, in the south-west of the Fezzan region (60 km north

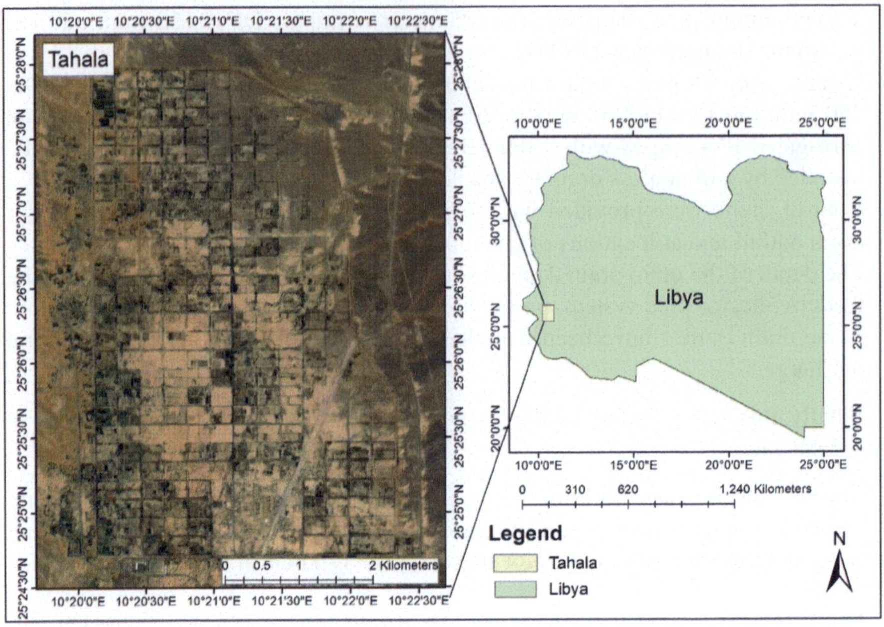

Fig. 6 Satellite image of the Tahala agricultural project and its location on the map of Libya

of Ghat city). The project was developed in 1974 as part of the agricultural development in Libya. In 1989, an Italian company was hired to install a drainage system and leach the accumulated salt in the root zone. In 1990, the drainage system was installed (subsurface drainage) (Hamilla and Alhadheri 2002). The drainage system was installed at depths ranging from 1.2 to 1.5 m and spacing from 12 to 43 m. The slope ranges from 1 to 2.4%, and the drain lengths were from 150 to 216 m. The diameters of the field drains were around 50 mm, discharging the excessive water to the sub-field drains (the diameters are 100 mm). After that, the excessive water is discharged using the main pipes with diameters starting from 200 to 600 mm (Hamilla et al. 2005).

3.3 *Agricultural Drainage and Reclamation of Saline Lands in the North-East Area*

3.3.1 Gwarsha Agricultural Project

The project is located south of Benghazi city. The project was established to use the treated wastewater for agricultural purposes, and the main crop grown in the project was forage (Forage was produced to feed the livestock). The project covers around 958 ha. The project's drainage system is subsurface drainage installed in depths ranging from 1 to 1.2 m. the spacing between drains was estimated to be about 30 m. However, the work on the project stopped in July 1993 when the wastewater treatment plant was out of service (GWA 2001).

4 Drainage Design Management (Design, Installation, Operation, and Maintenance)

As mentioned earlier, during the 1970s and 1980s, many drainage systems were designed and installed in some Libyan agricultural projects. The cost of installing a drainage system varied based on the nature of the problem and the geographical location. The average cost of installing subsurface drainage is around \$2.62 m^{-1} (Ellafi et al. 2021), which is expensive based on the average income in Libya. In a similar economic situation, other developing countries have made progress; for example, farmers in India and Pakistan contribute 15–20% of the operation and maintenance costs to overcome the high cost of installing and maintaining drainage systems. In Bangladesh, it is included in the water bill, while to install drainage systems, governments in Egypt, Bangladesh, India and Pakistan offer loans to farmers (Knegt 2000). Designing a drainage system for a specific location requires studying several factors affecting the drainage system. These factors are summarised as follows:

4.1 Geological and Soil Data

A collection of previous studies on the geology of the study area is needed. This will help identify the subsurface layers that impact the drainage system (e.g. the depth and thickness of the impermeable layer). Also, a geomorphology study is essential for drainage design. Drainage systems (commonly referred to as river systems in geomorphology) are the patterns formed by the streams, rivers, and lakes in a particular drainage basin. Drainage systems are influenced by the topography of the land, the gradient of the land, and whether hard or soft rocks dominate a specific location. Streams are frequently viewed as part of drainage basins by geomorphologists and hydrologists. This is the topographic area from which a stream gets through flow, runoff, and groundwater flow, which is its saturated equivalent. The topographic map's size and level of detail will determine how many, how big, and what kind of drainage basins are there (Pidwirny 2006).

Soil information is crucial for drainage design. These data can be obtained from previous soil surveys on a specific location. Accurately measuring soil data such as saturated hydraulic conductivity, soil moisture, and soil–water retention curve significantly impacts predicting the optimum drain spacing and depth. Also, these data would help to predict the irrigation water requirement for the desired crops (Ellafi et al. 2021). Measuring the electrical conductivity of soil extract is also important to identify areas affected by salinity. Also, estimating the exchangeable sodium percentage (ESP) and the sodium adsorption ratio (SAR) (calculated after determining Ca, Mg and Na concentrations in a saturation extract) is critical to understanding if the specific area is affected by sodicity or not.

4.2 Irrigation Water Quality and Quantity

Irrigation water contains salts, although in small yet considerable levels. It is developed from the dissolving or weathering of the soil's minerals, such as gypsum, and other slowly dissolved minerals. Wherever water is used, these salts are transported along with it. When water is used for irrigation, the salts are added along with it and stay in the soil while the water evaporates or is absorbed by the crop. The type and amount of dissolved salts in irrigation water may significantly impact its quality. Therefore, there is a need to evaluate the irrigation water quality and quantity for a specific project or location to identify the best possible management based on the quality and quantity of its water.

4.3 Cropping System and Irrigation Water Requirement

The proper and profitable crop rotation should be selected based on the location, climate, soil type and irrigation water quality. Many computer models can help to choose the suitable cropping system for a given site, such as CROPWAT (Smith 1992) and AQUACROP (Steduto et al. 2009) developed by FAO. These models can also help to determine the leaching requirement based on the soil information and the irrigation water quality.

4.4 Meteorological Data

Agro-climatic data such as temperature, relative humidity, wind speed, and precipitation are based on long-term observations of climatological data gathered across large areas and over extended periods. These data are used to study climatic variability and change and to find statistically significant links between climatological variables and soil and plant processes, animal, and pest development. However, in many areas in the world (especially in developing countries such as Libya), climatological data is unavailable daily or has missing values. Therefore, there is a need to explore the feasibility of using other alternative datasets, such as the reanalyses, including ERA5-Land (Muñoz-Sabater 2019) and MERRA-2 (Gelaro et al. 2017).

4.5 Drainage Models

Since the development of Darcy's Law of water flows in 1856, researchers and drainage system designers and operators have incorporated it into models that link the drain depth and spacing to the performance of drainage systems. The water movement in soil formula was developed by Colding (1872) to calculate water table depth in drained lands is also widely adopted (Rosbjerg 2020). Drainage models span a wide spectrum from complex guidelines that connect soil type to drainage requirements (Skaggs 1999) to drain spacing equations (Hooghoudt 1937), and computer models that simulate the daily performance of multi-component systems based on long-term weather data (Skaggs 1977; Skaggs et al. 2012a; Van Dam et al. 1997). Therefore, a drainage model can be defined as the compilation of guidelines, equations, or computer programs that are applied objectively to test the performance of a drainage system to be simulated (Skaggs 1999).

Under the above definition, analytical and numerical solutions (mathematical models) qualify as drainage models. Mathematical models have been applied to describe drainage events, typically over a limited period (Gurovich and Oyarce 2015). Mathematical drainage models such as Governing Equations and Boussinesq's Equation often consider a single process, such as drainage under a constant weather condition or water table drawdown for given initial and boundary conditions (Nieber and Feddes 1999; Youngs 1991). However, boundary conditions vary due to temporal and spatial changes in meteorological conditions, land cover, and land-use factors such as soil fertility, soil drainage, and slope. Simulation models predict the daily impact of these key factors' temporal and spatial variability on a drainage system's performance (Skaggs 1999).

Several simulation models such as DRAINMOD, SWAP and RZWQM2 have been applied to predict drainage systems' daily performance and monitor and control water table levels (Bennett et al. 2013; Skaggs et al. 2012a, 2012b). Many simulation models have been developed to describe a specific field condition, such as a shallow water table (Skaggs 1982), and the relationship between crop yield and drainage water management practices (Skaggs 1999). In humid areas, simulation models are applied to predict drainage outflow and water table fluctuations (Sanoja et al. 1990), crop production and nitrogen losses (Breve-Reyes et al. 1998), and nitrogen transformation and fate (Ale et al. 2012; Skaggs et al. 2012b). While in arid areas, simulation models such as DRAINMOD have been modified to characterize soil salinity and manage shallow water tables (Wahba and Christen 2006; Wahba et al. 2002). The downside of these simulation models is the substantial number of input parameters required to run them. These input parameters are expensive and hard to measure. Thus, there is a need to apply indirect methods to overcome the cost and time required to measure these input parameters and/or to fill in data gaps.

5 Recommendations

5.1 The Common Issues Related to Drainage Are Facing the Agricultural Projects in Libya

- The absence of proper and regular maintenance for the existing drainage systems within the agricultural projects.
- Unavailable drainage parts to replace any damaged pipe and equipment tools.
- The unmaintained drainage system failed to improve the yield, and the spread of Giant reed and Scirpus plants was noticed in many agricultural projects.
- Due to financial issues, unblocking the clogged drains and ditches is not possible.
- Limited support from the government to improve the existing drainage systems and design new ones within the agricultural projects.
- No commitment from farmers to stick with the recommended crop rotation.
- The roots of the trees grow in the drains and cause clogging.
- Even though the drainage systems designs are available, they are yet to be installed in some agricultural projects.
- Expanding the agricultural lands without extending and improving the drainage system.
- The absence of extension programs to educate farmers on the importance of agricultural drainage.

5.2 Actions Need to Be Taken to Improve the Drainage Within the Agricultural Projects

- Urgent moves are required to improve the existing drainage systems and install new ones for around 30,000 ha.
- Drainage water needs to be evaluated to assess the feasibility of reusing it for irrigation.
- Improve the irrigation management at the field level by providing farmers training and extension programs to improve water use efficiency.
- Work on establishing a local company or institution that focuses on installing, maintaining and improving the drainage systems in Libya.
- Apply technologies to improve the management of irrigation and drainage (e.g. drainage simulation models, geographical information systems, and cropping models.).
- Partner with drainage experts and researchers in Libya to develop training programs related to drainage design and management.

6 Conclusions

In Libya, drainage systems were crucial in maintaining the sustainability of agricultural projects. Agricultural drainage systems have been successfully installed in arid and semi-arid areas to control salinity and waterlogging issues. Even though drainage systems were installed in different regions in the country, salinity and waterlogging problems still exist in agricultural projects. Many reasons were behind the failure of drainage systems in Libya, including the absence of proper and regular maintenance for the existing drainage systems within the agricultural projects and the lack of extension programs to educate farmers on the importance of agricultural drainage. Also, the limited support from the decision-makers in the country, especially in installing drainage systems in areas that already have the design. GAID suggested many solutions to improve the drainage systems in Libya, such as establishing a local company or institution that focuses on installing, maintaining and improving the drainage systems and applying technologies to improve irrigation and drainage management. As the population in Libya continues to increase, and the demand for food and water resources will increase to fulfil the potential demand, an urgent move is needed to improve the existing drainage systems and install new ones for around 30,000 ha.

References

Al-Barouni (2013) The water situation in Libya. Tripoli, Libya

Al-Khoja, Hamilla M (2007) A report on irrigation and drainage in Al-Hamam and Al-Furjan agricultural projects general administration of irrigation and drainage. Tripoli, Libya

Ale S, Bowling L, Youssef M, Brouder S (2012) Evaluation of simulated strategies for reducing nitrate-nitrogen losses through subsurface drainage systems. J Environ Qual 41:217–228

Bennett SJ, Bishop TFA, Vervoort RW (2013) Using SWAP to quantify space and time related uncertainty in deep drainage model estimates: a case study from northern NSW, Australia. Agric Water Manag 130:142–153. https://doi.org/10.1016/j.agwat.2013.08.020

Breve-Reyes M, Skaggs RW, Parsons J, Gillian J (1998) Using the DRAINMOD-N model to study the effects of drainage system design and management on crop productivity and NO_3 losses in drainage water. Agric Water Manag 35:227–243

Chen W, Wang J, Liu Y, Jin M, Liang X, Wang Z, Ferré TPA (2021) Using bromide data tracer and HYDRUS-1D to estimate groundwater recharge and evapotranspiration under film-mulched drip irrigation in an arid Inland Basin, Northwest China. Hydrol Process 35:e14290. https://doi.org/10.1002/hyp.14290

Colding A (1872) Om lovene for Vandets Bevægelse i Jorden (On the laws for movement of water in soil). Vidensk. Selsk. Skr. 5:563–622

Cornelius-Brochier JV(1981) Wadi Al Shatti: drainage design. Tripoli, Libya

Danenco, (1980) Technical study for Jufra el Hamam project. Denmark, Copenhagen

Ellafi MA, Deeks LK, Simmons RW (2021) Application of artificial neural networks to the design of subsurface drainage systems in Libyan agricultural projects. J Hydrol Reg Stud 35:1–17. https://doi.org/10.1016/j.ejrh.2021.100832

Ellafi MA, Zurqani HA, Deeks LK, Simmons RW (2022) DRAINMOD applications to design drainage systems in Libya using soil salinity data predicted by GIS, remote sensing and artificial

neural networks. In: Zurqani HA (ed) Environmental applications of remote sensing and GIS in Libya. Springer International Publishing AG, Switzerland
FAO (2016) Country profile—Libya. Rome, Italy. https://doi.org/10.1787/9487b828-en
Gao X, Huo Z, Qu Z, Xu X, Huang G, Steenhuis TS (2017) Modeling contribution of shallow groundwater to evapotranspiration and yield of maize in an arid area. Sci Rep 7:1–13. https://doi.org/10.1038/srep43122
Gassman PW, Williams JR, Benson VW, Izaurralde RC, Hauck LM, Jones CA, Atwood JD, Kiniry JR, Flowers JD (2005) Historical development and applications of the EPIC and APEX models. Ames, Iowa. https://doi.org/10.13031/2013.17074
Gelaro R, McCarty W, Suárez MJ, Todling R, Molod A, Takacs L, Randles CA, Darmenov A, Bosilovich MG, Reichle R, Wargan K, Coy L, Cullather R, Draper C, Akella S, Buchard V, Conaty A, da Silva AM, Gu W, Kim G, Koster R, Lucchesi R, Merkova D, Nielsen JE, Partyka G, Pawson S, Putman W, Rienecker M, Schubert SD, Sienkiewicz M, Zhao B (2017) The modern-era retrospective analysis for research and applications, version 2 (MERRA-2). J Clim 30:5419–5454. https://doi.org/10.1175/JCLI-D-16-0758.1
Gupta GP, Prasher SO, Chieng ST, Mathur IN (1993) Application of DRAINMOD under semi-arid conditions. Agric Water Manag 24:63–80
Gurovich L, Oyarce P (2015) Modeling agricultural drainage hydraulic nets. Irrig Drain Syst Eng 4:1–9. https://doi.org/10.4172/2168-9768.1000149
GWA (2018) Folder reports of the general administration of irrigation and drainage. Tripoli, Libya
GWA (2007) A field visit to hammam agricultural project to investigate the Sokna flood. Tripoli, Libya
GWA (2005) The final report on agricultural drainage studies. Tripoli, Libya
GWA (2003) Report on the hammam agricultural project and the current drainage system in Jufra district. Tripoli, Libya
GWA (2001) Folder reports of the general administration of irrigation and drainage. Tripoli, Libya
Hamilla M (2018) Agricultural drainage in Libya. Tripoli, Libya
Hamilla M (2005) Summary of the reports of the most important experts, teams and committees that visited the Hammam and Al Furjan projects. Tripoli, Libya
Hamilla M, Alhadheri H (2002) Evaluations of the irrigation and drainage systems at Tahala agricultural project. Tripoli, Libya
Hamilla M, Shahoob A, Abia A, Alghati (2005) A report on the water situation in the Tahala agricultural settlement project. Tripoli, Libya
Hirekhan M, Gupta SK, Mishra KL (2007) Application of WaSim to assess performance of a subsurface drainage system under semi-arid monsoon climate. Agric Water Manag 88:224–234. https://doi.org/10.1016/j.agwat.2006.12.001
Holzmann-Wakuti (1974) Hydrogeology of the Joufrah project area. Tripoli, Libya
Hooghoudt S (1937) Contribution of the knowledge of some physical soil properties. NO 6. Detremination of the conductivity of soils of the second kind (in Dutch). Bijdragen tot de kennis Vaneenige natuurkundige Van den grand,6. Algemeene beschonwing Van het problem Van de d. The Hauge, Netherlands
Italconsult, (1976) Fezzan land reclamation project: Wadi-Shati Eshkeda area. Italy, Rome
Kale S (2012) Impact of drained and un-drained soil conditions on water table depths, soil salinity and crop yields. Afr J Agric Res 7:2935–2945. https://doi.org/10.5897/AJAR12.101
Kale S (2011) Field-evaluation of DRAINMOD-S for predicting soil and drainage water salinity under semi-arid conditions in Turkey. Spanish J Agric Res 9:1142–1155
Kandil H, Skaggs RW, Abdel Dayem S, Aiad Y (1995) DRAINMOD-S: Water management model for irrigated arid lands, crop yield and applications. Irrig Drain Syst 9:239–258
Kisekka I, Schlegel A, Vanderveer M, Ma L (2017) Optimizing preplant irrigation for maize under limited water in the high plains. Agric Water Manag 187:154–163
Knegt J (2000) Drainage in developing countries: a review of institutional arrangements. Wageningen, The Netherlands

Muftah K, Hamilla M, Shwtawi A, Aljaer A (2006) Waterlogging issues at Sokna Oasis. Tripoli, Libya

Muñoz-Sabater J (2019) ERA5-Land hourly data from 1981 to present [WWW document]. Copernicus Clim Chang Serv Clim Data Store

Nieber J, Feddes R (1999) Solutions for combined saturated and unsaturated flow. In: Skaggs R, van Schilfgaarde J (eds) Agricultural drainage, Crop Science Society of America, soil science society of America, Madison, Wisconsin, USA. American Society of Agronomy, pp 145–212

Pidwirny M (2006) The drainage basin concept. In: Fundemantals of physical geography

Qadir M, Boers TM, Schubert S, Ghafoor A, Murtaza G (2003) Agricultural water management in water-starved countries: challenges and opportunities. Agric Water Manag 62:165–185. https://doi.org/10.1016/S0378-3774(03)00146-X

Rosbjerg D (2020) Hydrology and beyond: the scientific work of August colding revisited. Hydrol Earth Syst Sci 1–15

Sanoja J, Kanwar RS, Melvin SW (1990) Comparison of simulated (DRAINMOD) and measured tile outflow and water table elevations from two field sites in Iowa. Trans Am Soc Agric Eng 33:827–833. https://doi.org/10.13031/2013.31407

Skaggs RW (1999) Drainage simulation models. In: Skaggs R, van Schilfgaarde J (eds) Agricultural drainage. American Society of Agronomy Inc., Madison, Wisconsin, USA, pp 469–500

Skaggs RW (1982) Field evaluation of a water management simulation model. Trans ASAE 25:666–674

Skaggs RW (1977) Evaluation of drainage-water table control systems using a water management model. In: St., A.P. 1–77., Joseph MA (eds) 3rd Natl. drainage symposium, pp 61–68

Skaggs RW, Youssef M, Chescheir G (2012a) A DRAINMOD-based method to estimate effects of drainage water management on annual nitrogen loss to surface water. Am Soc Agric Biol Eng 55:799–808

Skaggs RW, Youssef MA, Chescheir GM, Šimůnek J, van Genuchten MT, Šejna M, Arnold J, Moriasi D, Gassman P, Abbaspour K, White M, Srinivasan R, Santhi C, Harmel R, van Griensven A, Van Liew M, Kannan N, Jha M, Gowda P, Mulla D, Desmond E, Ward A, Moriasi D, Skaggs RW, Youssef MA, Chescheir GM, Šimůnek J, van Genuchten MT, Šejna M, Arnold J, Moriasi D, Gassman P, Abbaspour K, White M, Srinivasan R, Santhi C, Harmel R, van Griensven A, Van Liew M, Kannan N, Jha M, Gowda P, Mulla D, Desmond E, Ward A, Moriasi D (2012b) DRAINMOD: model use, calibration, and validation. Trans ASABE 55:1509–1522. https://doi.org/10.13031/2013.42259

Smedema LK, Abdel-Dayem S, Ochs WJ (2000) Drainage and agricultural development. Irrig Drain Syst 14:223–235. https://doi.org/10.1023/A:1026570823692

Smith M (1992) CROPWAT: a computer program for irrigation planning and management

Steduto P, Hsiao TC, Raes D, Fereres E (2009) AquaCrop—the FAO crop model to simulate yield response to water: I. concepts and underlying principles. Agron J 101:426–437. https://doi.org/10.2134/agronj2008.0139s

Taguta C (2017) Modeling subsurface drainage to control water-tables in selected agricultural lands LANDS in South Africa. Pan African University

van Aart R (1991) An evaluation of irrigation & land drainage problem in some settlement project in West Libya. Tripoli, Libya

van Alphen J, Kor G (1984) The final report Wadi Shatti drainage project & discussion on the water logging and salinity problems in the Jufra Al Hamam & Furgan irrigation project. Tripoli, Libya

Van Dam J, Huygen J, Wesseling J, Feddes R, Kabat P, Van Walsum PE, Groenendijk P, van Diepen C (1997) Theory of SWAP version 2.0. Simulation of water flow, solute transport and plant growth in the soil-water-atmosphere-plant environment. Wageningen.

Wahba MA, Christen E (2006) Modeling subsurface drainage for salt load management in Southeastern Australia. Irrig Drain Syst 20:267–282. https://doi.org/10.1007/s10795-006-9007-x

Wahba MAS, Kandil H, Gobran A (2002) Evaluation of DRAINMOD-S for simulating water table management under semi-arid conditions. Irrig Drain 51:213–226. https://doi.org/10.1002/ird.54

Youngs E (1991) A note on the power-law soils for deep. J Agric Eng Res 2:161–163

Mr. Mohamed A. Hamilla possesses a Bachelor's degree in Agricultural Sciences, specializing in water science and irrigation, earned from the Faculty of Agriculture at the University of Tripoli in 1984. With over thirty years of experience from 1984 to 2019, he was a prominent figure in the Department of Irrigation and Drainage at the General Water Authority (GWA). Over this period, he acquired extensive expertise in irrigation, drainage, soil management, and Geographic Information Systems (GIS) applications. He authored impactful reports and was pivotal in studies spanning 1984 to 2022. Mr. Hamilla collaborated extensively with the Food and Agriculture Organization (FAO) as a consultant on a project that harnessed the potential of treated wastewater and agricultural drainage water to advance agriculture in the Maghreb region, particularly Libya. Notably, he led the General Soil Administration from 2011 to 2013 and subsequently spearheaded information technology management within the Ministry of Water Resources from 2019 to 2023. Mohamed Hamilla's journey is marked by his unwavering commitment to advancing water management, soil conservation, and technological integration, both nationally and on an international scale.Mohamed possesses a Bachelor's degree in Agricultural Sciences, specializing in water science and irrigation, earned from the Faculty of Agriculture at the University of Tripoli in 1984. With over thirty years of experience from 1984 to 2019, he was a prominent figure in the Department of Irrigation and Drainage at the General Water Authority (GWA). Over this period, he acquired extensive expertise in irrigation, drainage, soil management, and Geographic Information Systems (GIS) applications. He authored impactful reports and was pivotal in studies spanning 1984 to 2022. Mr. Hamilla collaborated extensively with the Food and Agriculture Organization (FAO) as a consultant on a project that harnessed the potential of treated wastewater and agricultural drainage water to advance agriculture in the Maghreb region, particularly Libya. Notably, he led the General Soil Administration from 2011 to 2013 and subsequently spearheaded information technology management within the Ministry of Water Resources from 2019 to 2023. Mohamed Hamilla's journey is marked by his unwavering commitment to advancing water management, soil conservation, and technological integration, both nationally and on an international scale.

Dr. Murad Ellafi currently a Postdoctoral Research Associate at Northwest Research and Outreach Center—Crookston, Department of Soil, Water and Climate, University of Minnesota, achieved his Ph.D. at Cranfield University with a thesis dedicated to agricultural water management in arid and semi-arid regions. After completing his bachelor's degree in Agricultural Sciences, specializing in Soil and Water Science, at the University of Tripoli in 2010, he served as a Teaching Assistant from 2012 to 2015. Subsequently, Dr. Ellafi attained an M.Sc. in Environmental Water Management from Cranfield University in 2018, focusing his thesis on comprehending drain-aquifer interactions within low-lying agricultural drainage systems in eastern England. Dr. Ellafi is also a research visitor at Cranfield University since 2021 contributing lectures to M.Sc. students on watershed management. His present research pursuits encompass the application of machine learning for predictive drainage design parameters, evaluating controlled drainage effects on reducing nitrate and phosphorus loads, and assessing the influence of conservation practices on soil moisture dynamics.

Irrigation Development in Libya: Review and Assessment

Mohamed A. Hamilla and Murad Ellafi

Abstract This Chapter provides an overview of the history and perspective of irrigation development in Libya. The country faces water scarcity due to a shortage of water resources and increased demand for industrial, domestic, and agriculture expansion. The absence of maintenance programs for irrigation systems has resulted in low efficiency and early corrosion of tools. This issue has hindered the expansion of agriculture, leading to productivity losses and impacting development in irrigated lands. The Chapter evaluates the development of irrigation in Libya from the 1960s to the 2010s, including the challenges faced by agricultural irrigation in the country. Proposed solutions for improving irrigation water management and raising standards in this sector are presented, based on an extensive review of relevant studies, technical reports, and papers on water, irrigation, soil, and agriculture. The Chapter concludes that improving irrigation systems in Libya is crucial for agriculture and food security. To achieve this, necessary financial and technical resources must be provided, human capacities in irrigation must be developed, and technical information and specialized scientific references must be accessible to raise the level of performance and stay updated with scientific and technical developments in designing and evaluating irrigation networks. Finally, the Chapter presents several measures that may contribute to the advancement of the agricultural irrigation sector in Libya.

Keywords Irrigation development · Water scarcity · Challenges in irrigated lands · Agriculture and food security · Shortage of water resources · Agricultural productivity losses · Advancement of agricultural irrigation sector

---

M. A. Hamilla
Irrigation and Drainage Specialist, Tripoli 00218, Libya

M. Ellafi (✉)
Department of Soil, Water, and Climate, University of Minnesota, Twin Cities, MN, USA
e-mail: mellafi@umn.edu

Northwest Research and Outreach Center, University of Minnesota, Crookston, MN, USA

H. A. Zurqani (ed.), *Water Resources of Libya*, Springer Water,
https://doi.org/10.1007/978-3-031-80920-0_6

1 Introduction

Water scarcity in Libya is a significant problem resulting from a prolonged shortage of water resources. The issue has impacted the quantity and quality of some groundwater reservoirs due to continuous water drawdown levels with varying rates (Wheida and Verhoeven 2007). The increased demands on groundwater resources for industrial, domestic, and agricultural expansion, coupled with decreased rainfall over the entire country, have led to continuous seawater interference with varying rates along the coastline from one region to another (FAO 2016). This intrusion has caused the salt levels in coastal aquifers to exceed the recommended levels for freshwater declared by the Food and Agriculture Organization (FAO) (2016). Additionally, the nearby saltmarshes in the inner regions have increased salt quantity unless proper procedures are taken to stop groundwater overexploitation (Al-Barouni 2013; Ellafi et al. 2021, 2022). Furthermore, irrigation is a major concern in Libya due to the random increase in agriculture expansion over time, using systems such as furrow irrigation and sprinkler irrigation (GWA 2018).

The lack of maintenance programs for irrigation systems causes low efficiency and premature corrosion of tools. This issue hinders agricultural expansion, resulting in productivity losses and impeding development in irrigated lands (Hamilla 2002). To enhance agricultural production and support settlement projects, effective maintenance programs are critical, which must be tailored to specific irrigation systems and designed to address their inherent weaknesses. An exception to this is the Great Man-Made River's (GMMR) invested water projects, which have access to preventive maintenance programs. Without adequate maintenance, irrigation systems waste substantial amounts of water and rust before reaching their longevity, leading to reduced production in irrigated lands (Al-Barouni 2013).

This research evaluates the development of irrigation in Libya from the 1960s to the 2010s. It begins by defining and highlighting the importance of irrigation, as well as the water resources used for irrigation. An overview of agriculture and irrigated areas is also provided, along with a discussion of the advantages and disadvantages of various irrigation systems and their development stages over the last century. The article also discusses the challenges faced irrigated lands in Libya, including the policies implemented on irrigation networks in agricultural projects during this time period. Proposed solutions for improving irrigation water management and raising the standards in this sector are presented, based on an extensive review of relevant studies, technical reports, and papers on water, irrigation, soil, and agriculture. The research is the result of field visits to various agricultural regions during the random expansion of agriculture until the 2010s.

2 Irrigation Water Resources in Libya

Irrigation in simple words is the process of adding water to the soil to provide the necessary humidity for plant growth. Its purposes include adding water to the soil to ensure that plants are not dehydrated in the short term, cooling the soil and air to create an environment conducive to plant growth, decreasing the risk of frost, leaching soil to remove and reduce salt, and crumbling cohesive earth clods to facilitate tillage. In Libya, which is located in arid and semi-arid areas and experiences erratic rainfall, irrigation is of utmost importance. Rainfed agriculture dependents on weather factors, making irrigation processes crucial for crop success. Various methods and systems are utilized for irrigation, which will be discussed in detail later (Hamilla 2005).

In 2012, the total water withdrawal was estimated at 5830 million m^3 (FAO 2016). In which, 83% of it (around 4850 million m^3) for agriculture, 12% for domestic use (around 700 million m^3), and 5% for industries (around 280 million m^3) (FAO 2016). 95% of the water withdrawn is groundwater (around 5500 million m^3), 170 million m^3 is coming from surface water, desalinated water, and wastewater (CEDARE 2014). The National Strategy for Sustainable Development of 2008 (NSSD) determined that "sustainable" groundwater abstraction should not exceed 3650 million m^3 $year^{-1}$, despite the fact that only 650 million m^3 $year^{-1}$ comes from renewable groundwater, and 3000 million m^3 $year^{-1}$ comes from fossil water—from the Jefara plains (25 million m^3 $year^{-1}$), the Jabal al-Akhdar (25 million m^3 $year^{-1}$), the Kufra and Sarir (1300 million m^3 $year^{-1}$), the Hamada (150 million m^3 $year^{-1}$) and Murzuq (1500 million m^3 $year^{-1}$). Therefore, the present water extraction exceeds eight times the yearly renewable water resources since fossil groundwater is not included in the renewable water resources. In 2012, more than 50% of residential water supply originated from the Great Manmade River Project (Al-Barouni 2013).

2.1 *Groundwater*

Five large aquifers exist in Libya, with the first two linked to form the Western Aquifer (Table 1 illustrates the name of the aquifer, the total area within Libya, the quantity of renewable and non-renewable water, and the total dissolved). The shallow coastal of Al Jefara, Al Jabal al Akhdar, and, a portion of the Al Hamada aquifer, are naturally recharged by rainfall. Murzuq, Al Sarir-Kufra, and the remaining Al Hamada aquifer, which are situated south of 29° North latitude, belong to large sedimentary basins and contain fossil Quaternary water reserves. The Al Sarir-Kufra aquifer is a component of the Nubian Sandstone Aquifer System (NSAS), the world's biggest aquifer, spanning about 2 million km^2 across Libya, Chad, Sudan, and Egypt. The Al Hamada aquifer, on the other hand, is part of the North Western Saharan Aquifer System (NWSAS), which is shared by Algeria and Tunisia, and consists of the Terminal Complex sandstone and limestone aquifers and the Continental Interlayer sandstone aquifer (Wheida and Verhoeven 2007).

Table 1 The main aquifers in Libya, the amount of renewable and non-renewable water and their total dissolved

Aquifers	Areas (km^2)	Renewable (10^6 m^3)	Non-renewable (10^6 m^3)	Total dissolved (Mg l^{-1})
Al Hamada	215,000	250	150	1000–5000
Al jefara Plain	18,000	200	50	1000–5000
Al Jabal al Akhdar	145,000	200	50	1000–5000
Murzuq	350,000	–	1800	200–1500
Al Sarir-Kufra	700,000	–	1800	200–1500

Adopted from FAO (2016), and Wheida and Verhoeven (2007)

2.2 *Surface Water*

Surface water represents 2–3% of the total available water resources in Libya. This is due to the lack of rainfall, where more than 90% of the country receives less than 100 mm year^{-1}. However, it is estimated that the total runoff exceeds 250 million m^3 year^{-1} (GWA 2001). 18 dams have been constructed by the General Water Author (GWA) mostly on the northern part of the country with total capacity of 390 million m^3. However, it is estimated that the annual maximum water stored in these dams is around 60 million m^3. Table 2, shows the potential storage capacity and the existing storage of the dams in Libya.

2.3 *Wastewater*

The urban wastewater system received special attention in the 1960s. At that time, three wastewater treatment plants (WTPs) were established, as they were considered important infrastructure factors. Since then, the country has experienced sporadic economic and social development. This has resulted in an increase in water wastage and subsequently, a rise in wastewater production. Wastewater is now recognized as an unconventional renewable water resource. As a result, there has been a corresponding increase in the number of WTPs established to handle the growing volume of wastewater (Khajkhaj 2008). Wastewater can be considered to be an irrigation resource to reduce groundwater overexploitation that is used in agriculture. Documents about wastewater and WTP were reviewed by the general Water Authority and the Ministry of Housing and Utilities. The total number of WTPs is 80 stations, with varying sizes (Al-Barouni 2013). Around 20 stations are currently operational. 18 stations can be maintained, while 44 stations are consumed. The quantity of treated water in stations is 108350 m^3 per day. Treated water, when added to raw wasted water, represents 8.5% of the total (FAO 2016).

During the 1970s, an experiment was conducted to use treated wastewater for irrigation in Hadba Alkhadra (Tripoli) and Quwarsh (Benghazi) projects. The water

Table 2 The capacity and the existing storage of the Libyan dams as adopted from Al-Barouni (2013), GWA (2018)

Region	Underneath aquifer	Dam name	Location	Design capacity million m^3 year^{-1}	Existing capacity million m^3 year^{-1}
North east	Al Jabal al Akhdar	Wadi Qattara-1	Benghazi	135.0	12.0
		Wadi Qattara-2	Benghazi	1.5	0.5
		Marks	Ras-hlal	0.2	0.2
		Zara	Aloqurea	2.0	0.8
		Derna	Derna	1.2	1.0
		Abomansour	Derna	22.3	2.0
Central	Al Sarir-Kufra	Garif	Sirt	2.4	0.3
		Zhawia	Sirt	2.8	0.7
		Ziud	Sirt	2.6	0.5
		Benjuid	Benjuid	0.3	0.3
		Wadi Zgar	Jufra	3.7	0.2
North west	Al jefara Plain	Wadi Mjenean	Ben-Ghashir	58.0	10.0
		Wadi Ghan	Gheryan	30.0	11.0
		Wadi Zart	Rabta	8.6	4.5
	Al Hamada	Wadi Ekaam	Zliten	111.0	13.0
		Wadi Libda	Khomes	5.2	3.4
		Wadi Tibreat	Zliten	1.6	0.5
		Wadi Edkaar	Zliten	1.6	0.5
Total capacity				390	61.4

was used to irrigate crops such as wheat, barley, clover, corn, fruits, and vegetables, covering over 2000 ha in Hadba Alkhadra and 1000 ha in Quwarsh. These projects also provided feed for cattle farming. The use of treated wastewater in agricultural projects during that period brought benefits to the people in the area (Wheida and Verhoeven 2007).

3 Agricultural Projects and Irrigated Areas in Libya

Agriculture plays a crucial role in social and economic development as it is the primary source of food production and the foundation of stable communities, forming modern villages. Therefore, in the early 1960s, providing capabilities and financial assistance for the agriculture sector was a top priority. Complete plans for agricultural development were established in several areas, including Jaffara plain, central areas, Green Mountain, Fizzan, Kufra, and Assreer (GWA 2003, 2005; Hamilla et al. 2005).

Agricultural projects in Libya can be divided into three types. First one is irrigated projects. Such projects are available in areas that contain permanent ground and surface water gathered from the dams. The most modern irrigation systems and the most economical in using water were implemented such as the usual irrigation sprinkler one of them the pivot and the other is the localized irrigation, as well as the protected agriculture system inside plastic and glass greenhouses. The second type is rainfed projects. Rainfed projects were implemented in areas with heavy rainfall, where the annual rainfall rate is not less than 200 mm. These projects are considered resistant to drifting. In addition, terraces were established in the western region heights, such as Tarhouna heights, Misallata, Al Erban, and Ghiryan in the Jafarah plain region. Other projects included the Jarif valley project, Harawa, Sulan, and Hniwa in the central region, which are responsible for cultivating fruit and grain (GWA 2007). These are considered rainfall projects. Furthermore, a rangeland project was established in the central region to support livestock grazing. The third type is irrigated and rainfed projects. The projects implemented in areas with heavy rainfall, where the rainfall rate is not less than 200 mm per year, have access to both surface and groundwater resources. Some areas are designated for irrigation and used to produce vegetables and feed. Other areas receive rainfall in addition to irrigation, allowing for the cultivation of fruit and grain (Al-Barouni 2013).

4 Irrigation Systems Used in Libya

In Libya, a variety of irrigation systems are used to manage and distribute water resources for agricultural purposes. These include.

4.1 Sprinkler Irrigation System

This system is the most spread system at private farms contrary to other systems. Its advantages are easy to use, low maintenance price, portable and permanent or portable and permanent together (Lieth and Oki 2019). There are many types of sprinklers used in this system. One of the types is umbrella sprinkler, which is used at fruit and garden fields where it distributes water in the shape of an umbrella with a radius reaches 2–3 m, it covers a small area and it works under low pressure, one of its disadvantages it wastes surface torrent with high quantities and it has low efficiency. The second one is the rotating sprinkler which covers a high range of private farms, settlement projects, and common projects. There are several formations of these sprinklers that work according to the designed pressures, the areas they cover, and the sprinkling rates upon a specific schedule. The third one is the perforated pipes, which is used in nurseries, has low irrigation efficiency, and covers small areas (Lieth and Oki 2019).

4.2 Central Pivot Sprinkler Irrigation System

This sprinkler irrigation system is used in public and private projects for cultivating grains/seeds and feeds. It is used in southeast and west Libya and GMMR projects in central and west regions (GWA 2018). It is the second stage in the area it covers. The system has a long arm that moves by a source of power (electricity/diesel engine). It moves around the center of the circle that contains a source of water and power. The system can irrigate a large area up to 100 ha. It requires minimal labor and allows for multiple cultivation procedures simultaneously. There are no obstacles such as harvest, and spraying herbicide (O'Shaughnessy and Rush 2014).

4.3 Gun Sprinkler Irrigation System

A gun sprinkler irrigation system is a type of irrigation system that uses a sprinkler head mounted on a wheeled cart or tripod to distribute water over a large area. The sprinkler head resembles a gun, which is why it is called a gun sprinkler. The gun sprinkler system is designed to deliver large volumes of water at high pressure, which makes it ideal for irrigating large areas quickly, such as agricultural fields, sports fields, or parks. The sprinkler head can typically cover a circular area ranging from 20 m up to 90 m in diameter. This system is installed at Bai Al-Kabeer with dimensions of 100 × 100 m and at Green Feeds (Zawia) (Al-Khoja and Hamilla 2007; Hamilla 2005). One of the main advantages of a gun sprinkler system is its efficiency in covering large areas quickly and evenly. It is also relatively easy to move around and set up in different locations, making it a flexible option for irrigation systems. However, gun sprinkler systems can also be more expensive to operate and maintain compared to other types of irrigation systems, and they may not be suitable for areas with limited water resources or where water conservation is a concern (O'Shaughnessy and Rush 2014).

4.4 One-Sided Move Sprinkler Irrigation System (Linear)

The one-sided move sprinkler irrigation system, also known as a linear sprinkler irrigation system, is a type of mechanized irrigation system commonly used in agriculture. It involves a set of sprinklers mounted on a linear move irrigation system that moves back and forth across the field in a straight line. The system is typically powered by electric motors or diesel engines, and it can irrigate large areas of crops efficiently and effectively (O'Shaughnessy and Rush 2014). One-sided move sprinkler irrigation systems are particularly well-suited for rectangular or square fields with flat terrain, and they can be adapted for use in a wide range of crop types, including grains, fruits, and vegetables (GWA 2005).

4.5 Hanged Sprinkler Irrigation System

A type of mechanized sprinkler irrigation system that uses an arm with a set of sprinklers attached to it, which moves around a central point using a source of power. The system can cover an area of up to one hectare, and it is likely used for smaller-scale agricultural projects or in areas where the terrain does not allow for other types of irrigation systems. This system is implemented at the Annasr agricultural project (Tajora) (GWA 2001).

4.6 Localized Irrigation System

Localized irrigation systems, also known as micro-irrigation systems, are a type of irrigation system that applies water directly to the root zone of plants, typically through drip emitters or micro-sprinklers. This system is designed to provide precise and targeted watering, with water delivered slowly and evenly to the plants over a longer period of time. This helps to reduce water loss due to evaporation and runoff, making localized irrigation a more efficient method of watering crops (O'Shaughnessy and Rush 2014). It is commonly used in areas where water resources are scarce or limited, as it can help to conserve water while still providing adequate moisture for plant growth. Additionally, localized irrigation systems can be used in a variety of crop types, including fruits, vegetables, and flowers. Localized irrigation systems are widespread in many areas in the country and one of the disadvantages of using this system is blockage at the dripping holes with sand or salt deposits which demands continues maintenance (Hamilla 2005).

4.7 Surface Irrigation

Surface irrigation is a type of irrigation where water is distributed over the soil surface to infiltrate and provide moisture to the plant roots. There are three main types of surface irrigation systems:

1. Basin irrigation: In this method, the land is divided into rectangular or square basins surrounded by borders or ridges. Water is applied to each basin until it reaches the desired depth before being allowed to infiltrate the soil. Basing irrigation is an easy system to design and operate, low initial cost and suitable for a wide range of crops. However, it has a low water application efficiency, required a pre-work on the field to make sure the land is relatively flat and can cause soil erosion problems.
2. Furrow irrigation: In furrow irrigation, small channels or furrows are made in the soil between the crop rows. Water is then directed down the furrows and infiltrates the soil, providing moisture to the plant roots. The advantages of furrow surface

irrigation include reduced water loss due to runoff, lower water requirements compared to other surface irrigation methods, and reduced soil erosion. However, furrow surface irrigation also has some disadvantages, including difficulties in maintaining uniformity of water application, increased labor requirements for creating and maintaining furrows, and limited suitability for crops with widely spaced rows or uneven topography.

3. Border irrigation: This method is similar to basin irrigation, but the land is divided into long, narrow borders instead of square or rectangular basins. Water is applied to the top of the border and flows down to the bottom, infiltrating the soil and providing moisture to the plant roots. Border irrigation is an easy system to design and operate, suitable for a wide range of crops and has a slightly better water use efficiency compared to the basin irrigation. However, it can result in uneven water distribution if not properly designed and managed.

Basin and border irrigation were commonly used in Libya as historical irrigation methods, particularly for cultivating grain, vegetables, and trees along the coastline and oases, especially until the 1960s. Furrow irrigation was also a common method in many agricultural projects up to the 1990s (e.g. Tawergha, Algowarsha and Algatara projects) (GWA 2018).

5 Irrigation Development Stages in Libya

5.1 Old Irrigation Until 1960

Irrigation channels were found all over the oases and valleys, using shallow groundwater, and spring water for irrigation and animal watering. Coastal areas and oases have more cultivations due to the availability of groundwater that was accessed through traditional manual digging of shallow wells known as Arabian wells, with a depth not exceeding 40m and a diameter of 1.8–2 m. These wells were built and padded with stone brick or cement, with concrete rings reaching a height of 1 m and 7–10 cm thickness. Traditional methods of extracting water included animal power and mechanical tools such as a bucket were used. The bucket, (locally known as Alkum), was made of leather or rubber in a conical shape and contained a hole with a diameter of 50 cm and a size of 30–50 L (Tantawi 2000).

The extracted water was gathered in a surface tank (called Jabia-basins), then distributed through a water carrier to irrigation pools (Jadawel) that took time and effort to fill. Farmers' determination and experience have led to the development of surface irrigation systems such as shallow exposed water carriers, border irrigation systems, and small basins systems not exceeding 2×2 m. Through these basins, there are carriers to transfer water inside them in order to cultivate different types of grain, vegetables, and trees in their gardens, fields, and farms (Al-Khoja and Hamilla 2007).

Libyan farmers utilized spring water in various regions, including Arrabta, Ghadames, Sharshara, Ein Kaam, and other areas in the northern region such as Ein Annus, Mahrooqa, and Ein Biznty. Water distribution was carried out through a portion system with efficient counting methods, including tractor filling and the traditional rotation group irrigation system. The latter system involved open channels, as well as square or rectangular basins. Windmills were historically used in Libya for irrigation purposes. According to studies and reports, numerous windmills were installed on the coastline, and some still exist in locations such as Sidi Almasri (Tripoli), Khallut Juumaa (Janzor), Alloos (Khums), the Eastern region, and Mizda. However, due to the lack of regular maintenance, the requirement of large quantities of water, and the entry of diesel pumps into the market, windmills have become obsolete and neglected. As a result, the use of wind power and animal power for groundwater extraction has ceased (GWA 2005).

The utilization of diesel generators by farmers steadily increased after the discovery of oil and the availability of generators with various sizes and capacities. As a result, the use of human and animal power started to decline, and agricultural equipment and excavators became more accessible. With the proliferation of electricity and electronic pumps, water extraction also increased. Modern agricultural equipment, excavators, and pumps powered by diesel and electricity were used to facilitate groundwater wastage for cultivation. Additionally, various types of sprinkler irrigation systems were implemented, and hundreds of hectares were incorporated into permanent irrigation systems (GWA 2018).

5.2 *Irrigation Between 1961–1985*

During this stage, irrigation in Libya experienced significant changes and developments. The establishment of modern farms, settlement projects, and production led to a rapid growth in the irrigation sector, with the implementation of modern irrigation systems such as portable, pivot, linear, and gunner. The experience of using treated water in cultivation was also introduced, irrigating several areas with a total of 4000 ha. However, the increased use of irrigation also led to problems such as soil salinization, waterlogging, low efficiency of irrigation methods, and mismanagement of water within fields (GWA 2001).

To address these challenges, necessary legislation and rules were implemented to restrict water wastage and protect water resources. Water Rule (3) was declared in 1982 to promote the optimal and wise use of water resources and protect them from pollution. The establishment of the General Administration of Irrigation and Drainage within the General Water Authority in 1985 aimed to raise the efficiency of field irrigation and ensure continuous agricultural development. Overall, while irrigation in Libya underwent significant expansion and modernization, it also faced challenges that required attention and solutions (GWA 2018).

5.3 *Irrigation Between 1986–1990*

During this period, the general administration of irrigation and drainage was actively working to address the various problems and challenges facing the agriculture irrigation sector. Some of the main issues identified include poor selection of project locations, mismanagement of projects, and lack of regular maintenance of irrigation systems. In addition, there were challenges related to financial solvency, periodic analysis of irrigation water, and non-compliance with crop composition and agricultural rotations (GWA 2005; Hamilla 2005).

It is also interesting to note that wind energy was being explored as a way to withdraw groundwater for pastoral areas and that air fans were being installed as an experiment in various coastal areas. However, it seems that there were some challenges with the operational efficiency of these fans, which may have contributed to many of them going out of service. Overall, it appears that there was a lot of focus on improving the efficiency and effectiveness of the irrigation sector during this period (Al-Khoja and Hamilla 2007; Hamilla 2005).

5.4 *Irrigation Between 1991–2000*

The concept of establishing water cooperatives in the form of associations within settlement agricultural projects was first introduced in the mid-1990s. The aim was to provide guidance for water use in agriculture and enhance professional capacities, expertise, and management of water transmission and distribution networks to reduce water loss and excess consumption. To regulate water associations at the level of settlement projects, a proposal study was submitted based on Law (46/1971) and its executive regulations on agricultural cooperatives, which allows the establishment of agricultural cooperative societies (GWA 2007).

A technical committee, comprising the General Water Authority and the Department of Cooperatives in the agricultural sector, prepared a comprehensive vision of water associations to be applied in irrigated settlement projects. This vision included improving the extension programs and maintaining efficient irrigation systems. The role of farmers is essential in the real exploitation, management, and use of water equipment since they are the direct beneficiaries of this equipment and are responsible for managing it. The state's mission is to carry out maintenance, follow-up, and support. Therefore, the programs of water associations must be supported, which requires organizing extension programs in the short and medium term, to educate and improve the farmers' knowledge and awareness.

The technical committee recommended establishing water cooperative societies in all irrigated settlement projects on condition to be limited primarily to select agricultural projects in each region because of differences and variations of projects in terms of area, design, water facilities, and water requirements. Also, the technical committee recommended issuing a decision to declare water cooperative societies projects that are modernizing their infrastructure, such as the Kaam valley project, Nalut plane projects, Brak Iskda project, Alhmam project in Jufra, and Qasr Al-Akhyar project.

During this stage, great importance was given to the research and study aspect in developing appropriate solutions to suit local conditions to develop and improve drainage and irrigation systems to raise its efficiency and achieve the best management practices. The general administration of irrigation and drainage made a study about water requirements for important crops, which concluded by putting a valuation for crop water requirements. Close cooperation was launched during this stage between the general administration of irrigation and drainage with the Arab organization of agricultural development, where many studies about water management curricula, water policies, water rationalization, guidance, land use, irrigation methods use, and water association were prepared (GWA 1999).

Water investment devices were established for GMMR systems to organize, invest, and exploit water through policies, studies, plans, and through the prepared executive programs. The programs of investment devices for GMMR systems were divided according to its plans, such as supporting existing farms with irrigation water, reclamation of new areas under permanent irrigation systems, and reclamation of new areas as productive projects. The plans and programs for these devices have been changed to cope with the requirements of the stage.

5.5 *Irrigation Between 2001–2010*

During this phase, various programs were introduced to establish a strategic foundation for implementing initiatives aimed at promoting and rationalizing the use of agricultural irrigation water. These programs included the following:

1. Development of a manual for water and irrigation needs that covers a broad range of agricultural crops, fruit trees, and all regions.
2. Conducting a study on the establishment of water associations in settlement projects.
3. Studying and demarcating irrigation areas.
4. Evaluation of the effectiveness of the irrigation systems employed in Libya.
5. Study of the use of geographic information systems in managing irrigation water.
6. Establishment of an experimental farm for conducting irrigation and drainage experiments and research.
7. Rehabilitation of several existing agricultural projects through the preparation of technical specifications brochure.
8. Study of an irrigation equipment industry.
9. Implementation of the northern desert basin projects, with the first phase launched in 1999 and the second phase launched in early 2010. The primary objective of these projects is to use modern techniques for managing irrigated crops to generalize the applied experiments and transform them from an experimental

stage to a real applied stage. There was also discussion about the possibility of expanding the fourth stage to develop national policies, solve social problems, and provide economic capabilities (GWA 2003).

6 Policies and Strategies for Using Irrigation Water

The pursuit of achieving optimal self-sufficiency in agricultural production, particularly in vegetables, fruits, and meat, as well as attention to palm and olive trees, protection of natural resources, and optimization of exploitation through horizontal and vertical expansion programs have been the focal points of most Libyan policies. These policies also entail intensifying production elements, improving technical levels, and supporting infrastructure rehabilitation that facilitates the production process, including storage capacities, transportation, and distribution systems for irrigation water. To actualize these policies, three agricultural development plans were implemented within the comprehensive economic and social development plans, namely the Three-year plan 1973–1975, the First Five-Year Plan 1976–1980, and the Second Five-Year Plan 1980–1985, which have led to the establishment of numerous agricultural, settlement, and production projects that have garnered satisfactory results.

In the quest to manage water resources in Libya, the policies and strategies aim to mitigate the growing water deficit over time and reduce the deterioration of water quality in the affected areas, especially the coastal strip. These strategies revolve around reversing the consumption patterns of food in line with agricultural products that consume less water and have high nutritional value. Moreover, they aim to lay the groundwork for sustainable development that ensures the preservation of living standards and social and economic gains for the current generation, while securing the rights of future generations to a decent life by preserving and protecting water resources (Al-Barouni 2013).

7 Status of Irrigation Networks in Irrigated Projects

7.1 Private Sector

The efficiency of irrigation systems varies from weak to good depending on the system chosen by the farmer, including hand-transferred or fixed sprinkler irrigation systems, and drip irrigation systems. There are no restrictions on the choice of system. However, low efficiency is likely due to several reasons. First, the lack of regular maintenance may affect the performance of the system. Second, the absence of technical standards related to the design may disrupt hydraulic aspects. Third, even with good designs, non-compliance with the specifications specified for irrigation equipment may result in low efficiency. Fourth, the absence of standard technical specifications for irrigation equipment and systems according to each region due to the different nature of the soil, climate, and water capacity of these areas may affect the efficiency of the system. Finally, irrigation is often carried out randomly through the wrong convictions that more irrigation means more production (Hamilla 2005).

7.2 Public Sector

7.2.1 Settlement Projects

These projects were previously managed by specialized engineers who oversaw devices and units, including the Technical Affairs Unit. This unit supervised the irrigation programs within each project by linking a well or a group of wells to a group of farms directly or to the main tanks, which then pumped water to the farms' tanks for irrigation according to specified programs. However, farmers have recently been left to carry out all irrigation operations on their own. Despite these projects having high irrigation efficiency, problems have emerged over the past 25 years due to a lack of supervision and follow-up operations on irrigation programs, leading to a significant decrease in efficiency and a change in crop composition (Hamilla et al. 2005).

The lack of preventive maintenance programs has caused the waste of large amounts of water, which led to problems of waterlogging and salinization of some projects (Muftah et al. 2006). Non-compliance with the technical specifications of the irrigation network has also occurred when changing parts of pipes or some sprinklers or valves, leading to a breach of the hydraulic aspects and a lack of water access to some parts. Apparent shortcomings of supply have caused problems in providing the necessary operation and maintenance needs. The different forms, patterns, and types of irrigation systems have also caused problems due to the multinationalism of the companies designing and implementing these projects and the multiplicity of entities and supplied devices. Additionally, the absence of supervision over agricultural projects has led to most farmers drilling private wells, resulting in a change in the crop composition and expansion of the area (GWA 2005).

7.2.2 Productive Projects

Water amount and irrigation programs are supervised by project departments using modern high-efficiency irrigation machines, such as central pivot sprinkler irrigation system. However, field visits have revealed that some of these machines have low efficiency due to the lack of maintenance programs. Several attempts have been made to repair faults and replace worn-out irrigation units with new ones to improve water use efficiency and reduce excess loss (GWA 2018).

8 Conclusions and Recommendations

Irrigation in Libya is crucial in agriculture and food security. Improving irrigation systems in Libya will help to overcome the current issues such as salinity, water-logging and land productivity. It is essential to prioritize and support this sector by providing necessary financial and technical resources to ensure the efficient and sustainable use of these limited resources. Developing human capacities working in irrigation is also crucial, and continuous training and qualification, as well as access to technical information and specialized scientific references, can help raise the level of performance and stay updated with scientific and technical developments in designing and evaluating irrigation networks to improve field irrigation efficiency and management.

In summary, this Chapter represents a comprehensive summary of reports, studies, and papers prepared by experts in water, irrigation, soil, and agriculture to highlight the development stages of agricultural irrigation in Libya, including the challenges faced to optimize irrigation management. The success of agricultural irrigation development depends on restructuring this field and uniting efforts of relevant authorities to manage irrigated lands with good integrated management, as well as achieving development in the management of irrigated lands. Finally, the following measures may contribute to the advancement of the agricultural irrigation sector in Libya:

- Prioritizing spending budgets for short and long-term specialized training and extension programs.
- Evaluating agricultural irrigation facilities in the agricultural projects to improve water use efficiency.
- Introducing GIS and remote sensing programs for irrigation water management.
- Utilizing computer models specifically designed for the field of irrigation such as DRAINMOD, and AquaCrop.
- Benefiting from international and regional organizations working in the field of irrigation.
- Creating close cooperation between the Agricultural Research Center, water investment agencies of GMMR, and the General Authority for Water.
- Encouraging the establishment of specialized exhibitions for irrigation equipment, accessories, and pumps to stay updated with the rapid scientific developments in manufacturing and improvements and supporting the most water-saving irrigation devices.
- Improving irrigation management at the field level through better management programs, including training, and extensions.
- Continuously evaluating irrigated projects to obtain periodic information about the concerned party in this aspect.

References

Al-Barouni (2013) The water situation in Libya. Tripoli, Libya

Al-Khoja, Hamilla M (2007) A report on irrigation and drainage in Al-Hamam and Al-Furjan agricultural projects general administration of irrigation and drainage. Tripoli, Libya

CEDARE (2014) Libya water sector M&E rapid assessment report. Monitoring and evaluation for water in North Africa (MEWINA) project

Ellafi MA, Deeks LK, Simmons RW (2021) Application of artificial neural networks to the design of subsurface drainage systems in Libyan agricultural projects. J Hydrol Reg Stud 35:1–17. https://doi.org/10.1016/j.ejrh.2021.100832

Ellafi MA, Zurqani HA, Deeks LK, Simmons RW (2022) DRAINMOD applications to design drainage systems in Libya using soil salinity data predicted by GIS, remote sensing and artificial neural networks. In: Zurqani HA (ed) Environmental applications of remote sensing and GIS in Libya. Springer International Publishing AG, Switzerland

FAO (2016) Country profile—Libya. Rome, Italy. https://doi.org/10.1787/9487b828-en

GWA (1999) Water and irrigation requirements for the most important crops in Libya [unpublished report], Tripoli

GWA (2001) Folder reports of the general administration of irrigation and drainage. Tripoli, Libya

GWA (2003) Report on the hammam agricultural project and the current drainage system in Jufra district. Tripoli, Libya

GWA (2005) The final report on agricultural drainage studies. Tripoli, Libya

GWA (2007) A field visit to hammam agricultural project to investigate the Sokna flood. Tripoli, Libya

GWA (2018) Folder reports of the general administration of irrigation and drainage. Tripoli, Libya

Hamilla M (2002) Invistigate the use of automated irrigation to improve the water use effeciency in Libya. Khartoum, Sudan

Hamilla M (2005) Summary of the reports of the most important experts, teams and committees that visited the Hammam and Al Furjan projects. Tripoli, Libya

Hamilla M, Shahoob A, Abia A, Alghati (2005) A report on the water situation in the Tahala agricultural settlement project. Tripoli, Libya

Khajkhaj M (2008) Growth of small cities in Libya. Dar al-Sakia Publishing, Benghazi

Lieth JH, Oki LR (2019) Chapter 9—irrigation in soilless production. In: Raviv M, Lieth JH, Bar-Tal ABT-SC (Second E. (eds)) Elsevier, Boston, pp 381–423. https://doi.org/10.1016/B978-0-444-63696-6.00009-8

Muftah K, Hamilla M, Shwtawi A, Aljaer A (2006) Waterlogging issues at Sokna Oasis. Tripoli, Libya

O'Shaughnessy SA, Rush C (2014) Precision agriculture: irrigation. In: Van Alfen NKBT-E of A. and FS (ed). Academic Press, Oxford, pp 521–535. https://doi.org/10.1016/B978-0-444-52512-3.00235-7

Tantawi A (2000) Water resources in Libya. Egyptian Office for the Distribution of Publications, Cairo

Wheida E, Verhoeven R (2007) An alternative solution of the water shortage problem in Libya. Water Resour Manag 21:961–982. https://doi.org/10.1007/s11269-006-9067-6

Mr. Mohamed A. Hamilla possesses a Bachelor's degree in Agricultural Sciences, specializing in water science and irrigation, earned from the Faculty of Agriculture at the University of Tripoli in 1984. With over thirty years of experience from 1984 to 2019, he was a prominent figure in the Department of Irrigation and Drainage at the General Water Authority (GWA). Over this period, he acquired extensive expertise in irrigation, drainage, soil management, and Geographic Information Systems (GIS) applications. He authored impactful reports and was pivotal in studies spanning 1984–2022. Mr. Hamilla collaborated extensively with the Food and Agriculture Organization (FAO) as a consultant on a project that harnessed the potential of treated wastewater and agricultural drainage water to advance agriculture in the Maghreb region, particularly Libya. Notably, he led the General Soil Administration from 2011 to 2013 and subsequently spearheaded information technology management within the Ministry of Water Resources from 2019 to 2023. Mohamed Hamilla's journey is marked by his unwavering commitment to advancing water management, soil conservation, and technological integration, both nationally and on an international scale.

Dr. Murad Ellafi currently a Postdoctoral Research Associate at Northwest Research and Outreach Center—Crookston, Department of Soil, Water and Climate, University of Minnesota, achieved his Ph.D. at Cranfield University with a thesis dedicated to agricultural water management in arid and semi-arid regions. After completing his bachelor's degree in Agricultural Sciences, specializing in Soil and Water Science, at the University of Tripoli in 2010, he served as a Teaching Assistant from 2012 to 2015. Subsequently, Dr. Ellafi attained an MSc in Environmental Water Management from Cranfield University in 2018, focusing his thesis on comprehending drain-aquifer interactions within low-lying agricultural drainage systems in eastern England. Dr. Ellafi is also a research visitor at Cranfield University since 2021 contributing lectures to M.Sc. students on watershed management. His present research pursuits encompass the application of machine learning for predictive drainage design parameters, evaluating controlled drainage effects on reducing nitrate and phosphorus loads, and assessing the influence of conservation practices on soil moisture dynamics.

Irrigation Water Resources in Libya: Challenges and Potentials

Ali A. Shaki and Hamdi A. Zurqani

Abstract Due to the difficult climatic conditions in Libya, which include low rainfall and high rates of evaporation, water is a valuable and scarce resource. Water shortage in Libya has become a serious problem that requires efficient and sustainable methods of use and conservation. The primary objectives of this chapter are to: (1) explore the complex dynamics of water scarcity and irrigation practices in Libya, (2) discuss the importance of the agricultural economy in the country and its dependence on groundwater resources for irrigation due to the lack of surface water options; and (3) examine water quality considerations for irrigation and diverse irrigation practices in the region and highlighting issues such as water losses, soil salinity, and low crop productivity. The results confirm the value of effective water management in the country, which includes groundwater modeling, determining the amount of water needed for crops, and increasing irrigation system effectiveness. It is impossible to overestimate the significance of efficient water management in Libya, especially given the country's desert geography and few water supplies. Furthermore, the impact of climate change exacerbates the challenges faced by agricultural production in semi-arid regions such as Libya, and it is important to develop strategies to adapt to these changes.

Keywords North Africa · Water consumption · Irrigation practices · Irrigated agriculture · Crop water requirements · Murzuq basin aquifer system · Nubian sandstone aquifer system

---

A. A. Shaki
Department of Environmental Science, Faculty of Environment and Natural Resources, Wadi Al-Shatti University, Wadi Al-Shatti 00218, Libya
e-mail: a.shaki@wau.edu.ly

H. A. Zurqani (✉)
University of Arkansas Agricultural Experiment Station, Arkansas Forest Resources Center, College of Forestry, Agriculture, and Natural Resources, University of Arkansas, Monticello, AR, USA
e-mail: hzurqani@uark.edu; Zurqani@uamont.edu

H. A. Zurqani (ed.), *Water Resources of Libya*, Springer Water,
https://doi.org/10.1007/978-3-031-80920-0_7

1 Introduction

Libya is situated in North Africa and has a land area of approximately 1.759 million square kilometers (Zurqani 2022). The remaining 5% of the country comprises mountains, coastal regions, and oases, while about 95% is desert (Zurqani 2021). Only 1.2% of the land area is considered arable, or around 2.2 million hectares. According to estimates, Libya has an irrigation potential of 750,000 ha (FAO 1997). However, the usage of fossil water would be primarily necessary for the development of this potential. It is predicted that a maximum of 40,000 ha might be irrigated in the coastal regions when renewable water resources are taken into account (FAO 1997). Due to the country's dry environment and limited water supplies, irrigation development in Libya is complicated. Fossil water consumption for irrigation can result in water shortages and aquifer depletion. Thus, the development of irrigation in Libya will require careful planning and management of water resources, as well as cutting-edge technologies such as drip irrigation to improve irrigation in Libya.

Water scarcity in arid and semi-arid regions is mainly attributed to limited precipitation and harsh climatic conditions, increasing evaporation rates, and subsequent water losses. Countries such as Libya have substantial challenges due to their limited annual rain since around 93% of their area receives less than 200 mm of rainfall annually (IMB 1977). This implies that only 1% of the country's land area experiences sufficient precipitation levels to sustain agricultural activities. In addition to its limited precipitation, Libya has high daily evaporation rates exceeding 6.8 mm on average (IMB 1977). The daily evaporation rate is 60 times greater than its average precipitation, leading to unsustainable use of surface water resources. Given the limited rainfall and high evaporation rates in the region, the agricultural sector faces significant challenges in ensuring sustainable water usage. This sector is highly dependent on water, with 85% of total water consumption in the country (NSIWRM 1999). This indicates the significant role agriculture plays in the country's water resources management. In addition to its limited precipitation, Libya has high daily evaporation rates exceeding 6.8 mm on average (IMB 1977).

The agricultural terrain in Libya is limited by the environment and labor shortages. It consists of 13.3 million hectares of permanent pastures, 1.8 million hectares of annual crops, and 0.3 million hectares of permanent crops (FAO 1995b). According to (LIB/00/004 2018), the agricultural sector contributes less than 9% of the gross domestic product (GDP) and employs approximately 5% of the labor force. Private farms produce the majority of output (FAO 1995b). At the same time, government irrigation projects provide support for grain and forage production in desert regions (Zurqani et al. 2019) (Fig. 1). However, the agricultural industry encounters remarkable challenges, including issues related to water scarcity, soil salinity concerns, and the prevalence of pests and diseases. The implementation of irrigation systems in Libya has the potential to mitigate the difficulties mentioned earlier. However, it is important to prioritize sustainability and avoid exacerbating the depletion of the country's water supplies. Groundwater irrigation constitutes nearly 99% of the irrigated land in Libya. Meanwhile, treated wastewater and surface water resources are used to irrigate the remaining one percent of the land. The exploitation of groundwater for agricultural irrigation has generated concerns regarding the scarcity of water supplies and the steady depletion of subterranean aquifers.

Moreover, irrigated agriculture frequently demonstrates inefficiency, as a substantial amount of the water is lost because of evaporation or runoff (Shaki 2012). To exacerbate the situation, the population of Libya is concentrated in the coastal region, where groundwater resources are insufficient (MAWINA 2013). This means that the available water is being used to meet the needs of a small number of people, while the vast majority of the country's population does not have access to enough water (MAWINA 2013). The government faces the ever-present challenge of balancing the water demand and unequal water access with preserving the country's groundwater resources (MAWINA 2013). The situation in Libya is a stark reminder of the challenges that arid and semi-arid countries face in terms of water security. These countries need to find ways to develop and manage their water resources more sustainably, or they will face severe water shortages in the future (MAWINA 2013).

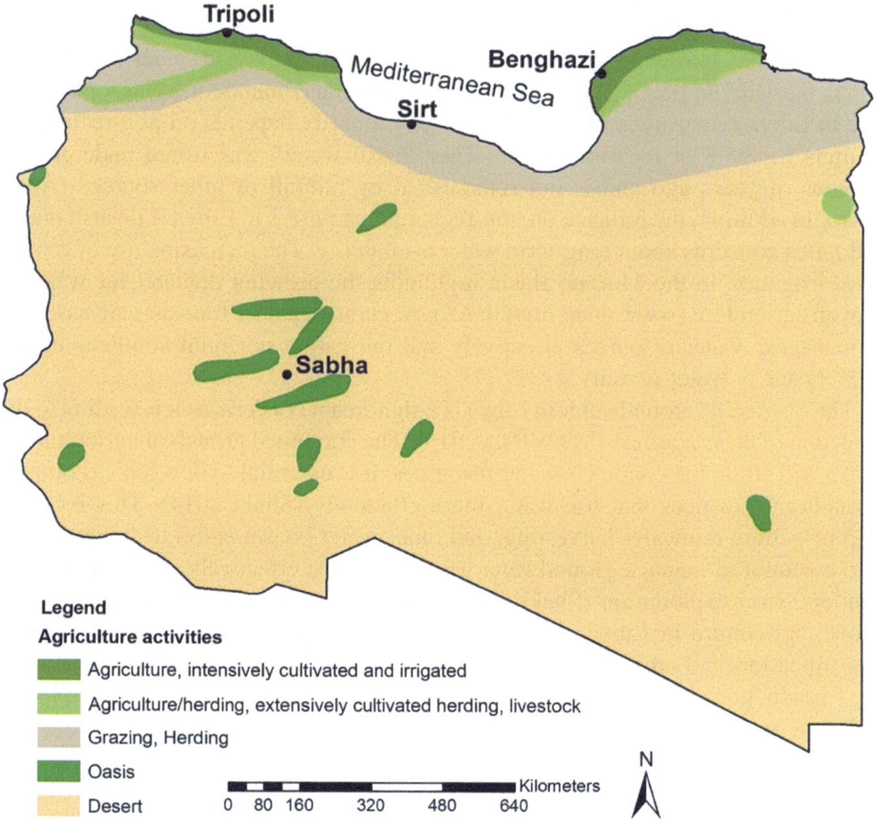

Fig. 1 Locations of agricultural activities in Libya (adopted from: Zurqani et al. 2019)

The scarcity of surface water resources for agricultural use in Libya presents a real challenge due to its ephemeral rivers (NSIWRM 1999). The utility of treated wastewater is considered a more sustainable option, but its availability curtails its usage. The amount of water used for agriculture in 2012 is estimated at 4850 million cubic meter (mm^3) per year, but the amount withdrawn is unknown due to inadequate monitoring infrastructure and unregulated groundwater usage (NSIWRM 1999). This lack of information and regulation adds to the uncertainty surrounding water resources in Libya and makes it difficult to manage and allocate this precious resource effectively. To address the challenges of limited surface water and unregulated groundwater usage, Libya must improve its monitoring infrastructure and establish regulations for water usage in agriculture. In addition to addressing the challenges of limited surface water and unregulated groundwater usage, it is also critical for Libya to invest in water conservation measures such as improving irrigation efficiency, implementing water-saving technologies, and promoting sustainable farming practices.

The satellite images in (Fig. 2) demonstrate a significant increase in center-pivot irrigation in the Murzuq Basin in southeastern Libya between 1987 and 2010 (www.unep.org/GEAS). This irrigation method involves rotating pipes to water crops and is highly efficient. However, it also requires a lot of water. The rise in center-pivot irrigation in the Murzuq Basin clearly indicates the growing demand for water for agriculture in Libya (www.unep.org/GEAS). Libya primarily depends on non-recharging aquifers for 95% of its water needs. This "fossil water" was stored underground millions of years ago and is not replenished by rainfall or other sources (Alker 2008). In addition, the reliance on non-recharging aquifers in Libya is unsustainable and raises concerns about long-term water availability. The increasing use of center-pivot irrigation in the Murzuq Basin highlights the growing demand for water in Libyan agriculture (www.unep.org/GEAS). Accurate weather forecasts are essential in managing water resources effectively and mitigating potential conflicts among farmers due to water scarcity.

The overuse of groundwater in Libya is a significant concern, as it is leading to the depletion of these aquifers (MAWINA 2013). The continued growth of agriculture in Libya will strain the country's water resources. It is essential to develop sustainable agricultural practices that use water more efficiently (Shaki 2014). This includes drip irrigation, rainwater harvesting, and other water conservation techniques. It is also essential to manage groundwater resources more effectively and reduce these aquifers' over-exploitation (Shaki 2006). In addition to the challenges mentioned above, agriculture in Libya also faces other challenges, such as climate change, desertification, and salinization, all of which must be addressed to ensure sustainable food production in the country (MAWINA 2013). The development of irrigation in Libya needs a careful balance between increasing agricultural production and protecting the country's water resources. Thus, innovative technologies such as drip irrigation can improve irrigation efficiency and reduce water usage.

In addition, irrigated agriculture is often inefficient, with much of the water being lost through evaporation or runoff (Shaki 2014). To exacerbate the situation, the population of Libya is concentrated in the coastal region, where there is less groundwater.

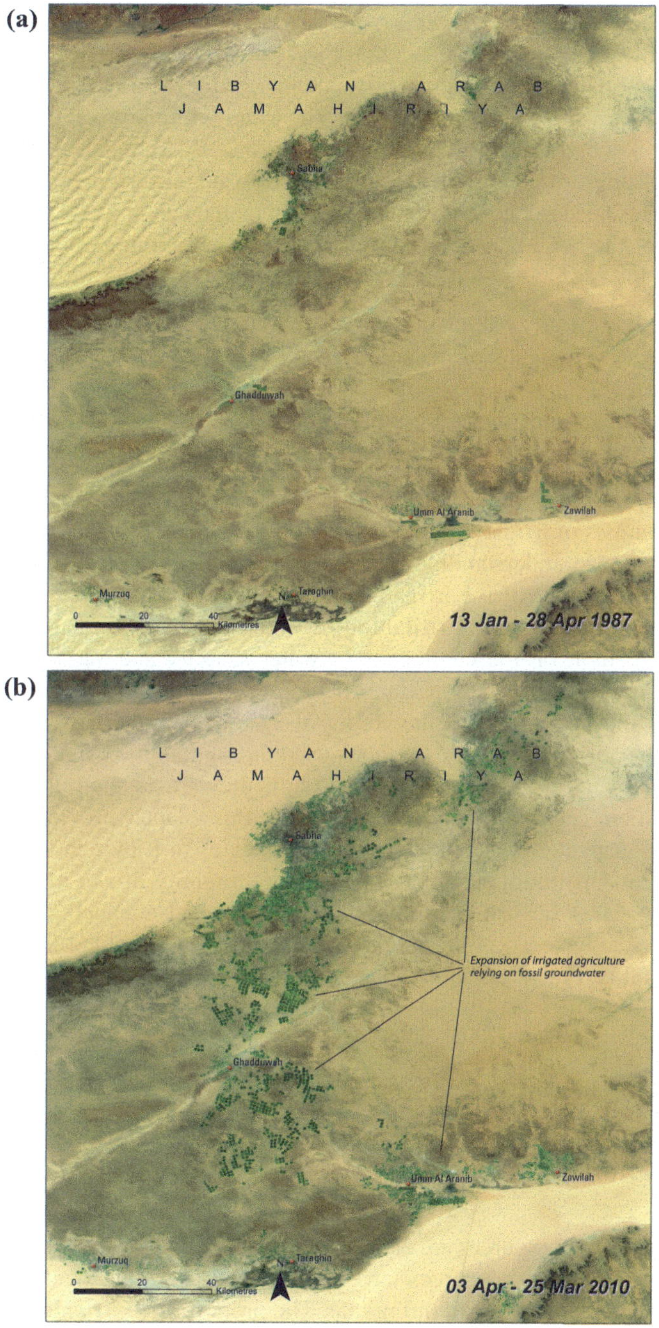

Fig. 2 Illustrate the significant increase in center-pivot irrigation on the Murzuq Basin in south-eastern Libya between 1987 and 2010, **a** satellite imagery from 1987 shows some irrigation in the Murzuq Basin (www.unep.org/GEAS), and **b** green squares and circles of irrigated crops stretch across the Murzuq Basin in this March 2010 satellite image (www.unep.org/GEAS)

This means that the available water is being used to meet the needs of a small number of people while the majority of the country's population does not have enough water (NSIWRM 1999). This is the most critical challenge for the government, as it needs to find ways to ensure everyone has access to water while protecting the country's groundwater resources. The situation in Libya is a stark reminder of the challenges that arid and semi-arid countries face in terms of water security (MAWINA 2013). These countries need to find ways to develop and manage their water resources more sustainably, or they will face severe water shortages in the future. Some of the ways to do this include:

- Investing in rainwater harvesting and other water conservation measures.
- Improving the efficiency of irrigation systems.
- Developing alternative water sources, such as desalination.
- Promoting water-sensitive agriculture.
- Educating the public about the importance of water conservation.

In this chapter, we thoroughly explore the key themes pivotal in understanding the complex dynamics of irrigation practices in Libya. It covers various topics essential to a comprehensive understanding of irrigation in Libya. We begin by conducting a comprehensive study of the irrigation water resources in the region, examining the sources, availability, and distribution of water that underpin agricultural activities, shedding light on the challenges of water scarcity and variability. We then evaluate the quality of irrigation water, which profoundly influences crop health, soil fertility, and the overall sustainability of agricultural industries. Furthermore, we explore the various irrigation methods used throughout the country, considering both traditional and modern techniques and how they are adapted to local environmental and socio-economic conditions. Overall, this chapter provides an in-depth exploration of irrigation water management strategies in Libya, including governmental policies, institutional frameworks, and community-driven initiatives aimed at optimizing water utilization, mitigating wastage, and enhancing the resilience of Libyan agriculture. The topics covered in this chapter include irrigation water resources, quality, practices, and management.

2 Irrigation Resource Management Overview

2.1 Irrigation Water Resources

In Libya, most water resources are classified as non-renewable and are often referred to as fossil water. These resources have accumulated over thousands of years and are now being depleted at an alarming rate. The proportion of irrigation water withdrawals derived from groundwater aquifers accounts for 98.7% of the total area, while surface water sources contribute to irrigating 0.65% (LIB/00/004 2018). In addition, wastewater is used for irrigation in 0.65% of the land (LIB/00/004 2018).

The overuse of groundwater in Libya for agricultural purposes is a significant concern that is leading to the depletion of underground water and causing salinity and sea-water penetration into coastal aquifers (LIB/00/004 2018). This alarming depletion of underground water in Libya due to agricultural activities calls for urgent measures to ensure the sustainable use of these limited resources. Figure 3 illustrates the statistics on irrigation water resources in Libya.

As stated earlier, a significant portion of agricultural practices rely on groundwater resources, particularly in the country's southwestern region, where the Murzuq Basin contributes the most important volume of water. The water withdrawal statistics for agricultural use across many basins from 1978 to 2010 are presented in (Table 1). The data shows the variations in water withdrawals across this time frame. The water withdrawals in the Hamada basin increased from 173.1 million cubic meters in 1978 to 241.2 million cubic meters in 1984 and remained constant at 350 million cubic meters from 1984 to 2010. The Al-Jabal Al-Akhdar basin also experienced an increase in water volume from 79.5 million cubic meters in 1978 to 150.5 million cubic meters in 1984. Then it remained stable at around 480 million cubic meters from 1995 forward. The Kufra and Sarir basin expanded significantly from 216.5 million cubic meters in 1978 to a maximum of 905 million cubic meters in 2010. In 1978, the Al-Jifarah basin had 435 million cubic meters of water, which rose to 1110 million cubic meters in 2010. The volume of the Murzuk basin increased significantly from 372.5 million cubic meters in 1978 to 2077 million cubic meters in 2010. To summarize, water consumption trends changed substantially from 1978 to 2010, with the total volume of water withdrawals from all basins increasing from 1276.6 million cubic meters to 4922 million cubic meters.

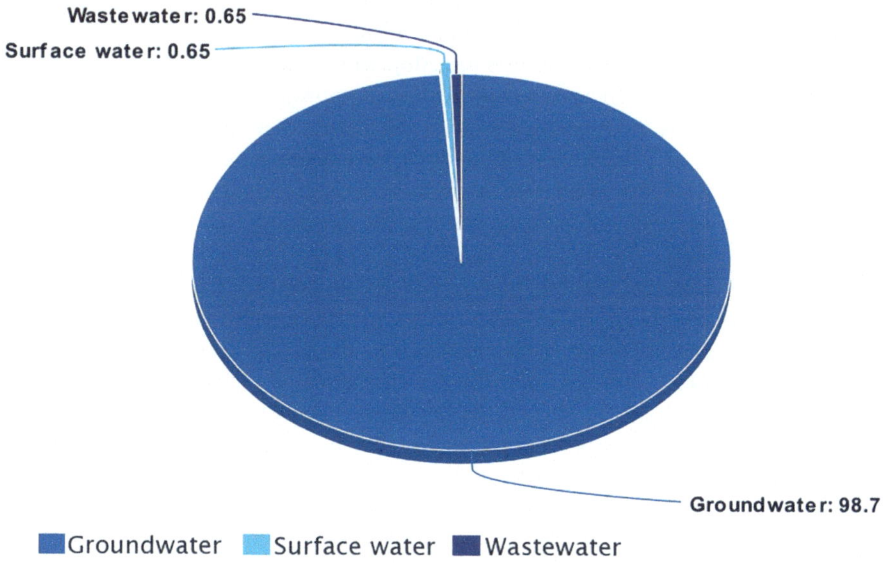

Fig. 3 Irrigation water resources in Libya (adopted from: LIB/00/004 2018)

Table 1 Water withdrawals for agricultural use in million cubic meter (1978–2010)

Basin name	1978	1984	1995	2005	2010
Hamada	173.1	241.2	260	350	350
Jabal Al Akhdar	79.5	150.5	469	480	480
Kufra and Sarir	216.5	335	335	650	905
Jifarah	435	500	965	1110	1110
Murzuk	372.5	551	751	1848	2077
Total	1276.6	1777.7	2780	4438	4922

Source CEDARE (2014)

Large-scale changes impacting groundwater resources have already been observed in this region, with a regional study in Morocco revealing a drop in groundwater levels in several basins (Fleury et al. 2023). This drop in groundwater levels has caused the water table to disappear in areas with shallow aquifers. These changes in groundwater levels can be attributed to the shift in agricultural practices from rainfed farming to groundwater irrigation. The excessive use of groundwater resources poses severe consequences to the environment and livelihoods (Liu et al. 2016).

2.2 *Irrigation Water Quality*

The irrigation water quality in Libya varies depending on its source and geographic location. Groundwater is the primary water source, but it is vulnerable to contamination from industrial and agricultural activities, resulting in elevated salt levels and pollution. The Libyan government has adopted several initiatives to monitor and enhance water quality, such as initiating desalination projects and promoting treated wastewater utilization for farming purposes. However, there are still challenges to overcome, especially in remote areas with limited access to clean drinking water. In Libya, there is a collective effort to improve irrigation water quality. However, sustainable methods and increased investment are necessary to address the existing challenges effectively.

Many studies have documented that irrigation water quality in Libya is a growing concern. A study conducted by Shaki and Adeloye (2006) on the Gadowa agriculture project found that most water wells belonged to the C1-S1 (low salinity and low sodium content) class, which is suitable for irrigation on almost all types of soil. Only two samples belonged to the C2-S1 (medium salinity and low sodium content) class, which is also suitable for most plants provided moderate leaching takes place or for plants with moderate salinity tolerance without considerable practices for salinity control (Fig. 4). Furthermore, Shaki and Zaid (2009) conducted a study on the Tsawa Irrigation project using the classification of the American Salinity Laboratory. The results showed that all project wells belonged to class (C1-S1), meaning the water can irrigate all crops without any precautions. According to the FAO Index, only well

No. 15 belonged to class (C2-S1), which can be used in most crops with additional water quantities. There were no issues with using all water for irrigation in terms of salinity, permeability, and toxicity.

Also, in a study by Shaki and Abdalla (2021) in Wadi Shatti, samples were collected from 27 wells in various villages. The classification of the American Salinity Laboratory was employed to analyze the water quality of these wells. The results indicated that all wells in the region were classified as belonging to the (C3 S1) category, representing high-salinity low-sodium water. This classification suggests that the water can be used for irrigation in soils with medium to reasonable basic infiltration rates. However, it is essential to implement a well-structured washing system to prevent salt accumulation in the soil. The likelihood of increased adsorbed sodium in the soil is very low. According to the FAO Index, using this water for irrigation presents many problems. All water samples were classified as having a second-degree salinity problem, meaning there is a high possibility of increased salinity when using water. On the other hand, all samples were in the first category regarding affecting permeability. However, there was a problem with chloride, bicarbonate, and sodium in terms of toxicity index.

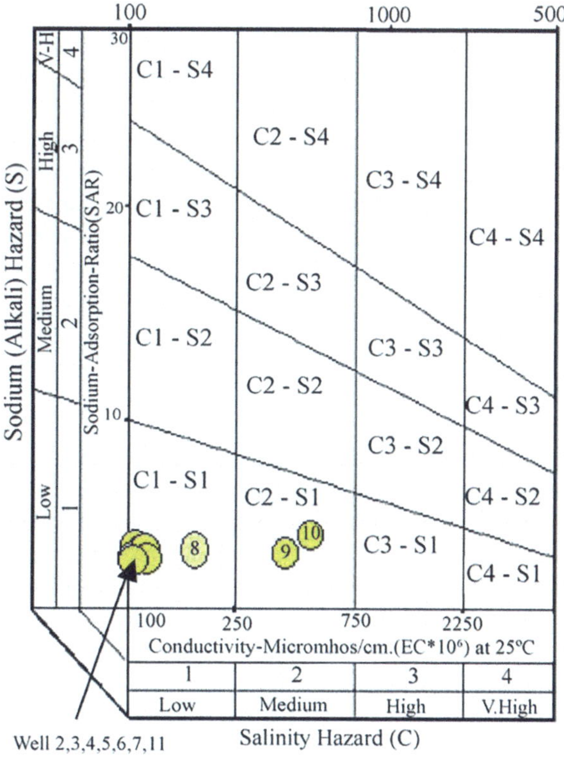

Fig. 4 Classification of ground water in Gadowa agriculture project (*Source* Shaki and Adeloye 2006)

2.3 Irrigation Practices

Irrigation practices in Libya heavily rely on underground sources such as wells and boreholes, as the country has limited surface water resources. Traditional methods such as flood and furrow irrigation are still widely used, although there is a growing shift towards more efficient and sustainable practices like drip and sprinkler irrigation. The government has initiated projects to improve water efficiency and enhance irrigation infrastructure to meet the increasing demand for agricultural production. However, challenges such as water scarcity, high evaporation rates, and limited access to technology and resources pose obstacles to implementing advanced irrigation techniques in Libya. This region has one of the largest groundwater basins in the country, the Murzuq Basin. However, 100% of the irrigation practices in the southwest of Libya occur in an area with high temperatures, evapotranspiration, and no rainfall. To mitigate the challenges caused by these conditions, efficient irrigation techniques and water management strategies should be implemented.

2.3.1 Irrigation Patterns

There are three main types of agricultural projects in Libya: Public Agriculture Projects managed by the state (over 1000 ha with mean production of cereal), Settlement Agricultural Projects managed by farmers (around 10 ha with mixed production including cereal, vegetables, and fruits), and Private Sector projects (ranging from 0.5 to 100 ha with mean production of cereal). The diversity of these agricultural projects in Libya reflects the different scales and management approaches contributing to the country's overall food production. However, the low share of farm workers in Libya suggests that the agricultural sector has a relatively small impact on the national economy compared to other countries in the region.

2.3.2 Irrigation Systems

Libya faces several challenges in implementing effective irrigation systems. The country's arid climate, lack of water resources, and political instability have hindered the development of proper irrigation infrastructure. However, investments in modern irrigation technologies and techniques have been made to maximize water usage and improve agricultural productivity. Various methods such as flood, sprinkler, and drip irrigation are utilized to optimize water usage. These systems help efficiently deliver water to crops, minimize wastage, and maximize crop yield. Surface irrigation systems, such as sprinklers, solid sets, center-pivot, and localized systems, are commonly used in older farms. While center-pivot irrigation is becoming increasingly popular nowadays (Figs. 5, 6, 7 and 8).

Fig. 5 Surface irrigation systems (field photo taken by: Shaki 2010)

Fig. 6 Sprinkler irrigation systems (field photo taken by: Shaki 2010)

While conducting field visits to different agricultural sites in Libya, there are some irrigation practices that can be addressed. These practices include a lack of operation experience, huge loss of water, Increasing soil salinity, and Low crop productivity.

Fig. 7 Localized irrigation systems (field photo taken by: Shaki 2010)

Fig. 8 Center-pivot irrigation systems (field photo taken by: Shaki 2010)

Lack of Operation Experience

Figure 9 illustrates the lack of experience operating Center-pivot Irrigation systems in the private sector in Libya, which has become increasingly popular in recent years. This lack of experience in operating Center-pivot Irrigation systems in the private sector in Libya highlights the importance of educational and training programs to

Fig. 9 Canceling one by one nozzle, in the area between Gadowa and Murzuq (field photo taken by: Shaki 2010)

ensure the successful implementation and management of these systems. In addition, it is essential for the government and relevant stakeholders in Libya to collaborate in developing policies and regulations that support the adoption and operation of Center-pivot Irrigation systems in the private sector.

High Water Losses

As shown in (Figs. 10 and 11), many of the fields used for cultivation were not leveled beforehand, and there was no consideration for soil properties related to irrigation design, such as basic infiltration rate (Shaki 2021). When the irrigation water addition rate exceeded the basic infiltration rate, runoff occurred (Fig. 10). One other issue observed was the leakage coming from the pipes (Fig. 11).

Low Crop Productivity

One of the main reasons for low crop productivity in Libya is the absence of proper irrigation infrastructure. The Al-Al-Arial agriculture project witnessed relatively low production rates for wheat and barley, even though they were the highest in their respective years. The wheat production rate was 3.6 tons per hectare in 2003/2004, while barley had the same production rate in 2005/2006. However, this rate is still considered low, indicating the urgent need for improved irrigation infrastructure to enhance crop yields in Libya.

Fig. 10 Irrigation water runoff and gathering outside the field at Gadowa area (field photo taken by: Shaki 2010)

Fig. 11 Leakage from the pipe (field photo taken by: Shaki 2010)

Increasing Soil Salinity

Soil salinity accumulation is one of the major problems in Libyans irrigation, causing disruptions in plant growth and reducing crop productivity. This problem increases by the extensive use of saline water for irrigation and the application of fertilizers,

Fig. 12 Salinity of soil at Gadowa project (field photo taken by: Shaki 2010)

Fig. 13 Salinity of the soil in private farm at Sebha area (field photo taken by: Shaki 2010)

which create unsuitable conditions for crop production. Long-term use of irrigation water without the addition of leaching water can increase soil salinity. Figures 12 and 13 show a case of soil salinity issues at the Gadowa project.

Fig. 14 Draining water from the field in Arial project Wadi al-Shati (field photo taken by: Shaki 2010)

Low Soil Holding Capacity

In Libya, farmers face difficulties providing adequate moisture to their crops due to the low soil holding capacity, particularly in sandy fields where water is drained quickly (Fig. 14). This makes it difficult for the crops to get sufficient water and nutrients, leading to decreased crop yields and potential agricultural losses. Farmers may need to explore strategies such as soil conditioning and improving irrigation systems to address this issue and ensure better crop growth and farming sustainability in the region.

Applying Deference Water Requirement

Irrigation in Libya requires applying sufficient water to meet crop water requirements. However, these requirements can vary from year to year. For instance, the Dabawat agriculture project found that wheat crops needed 13,441 m^3/ha/season in 2008–2009, compared to 10,083 m^3/ha/season in 2005–2006. Therefore, it is important to monitor and adjust irrigation practices accordingly to ensure successful crop growth in the region.

2.4 Irrigation Water Management

Agricultural irrigation involves providing plants with necessary water during different growth periods. Irrigation efficiency requires a good understanding of soil sciences, agriculture, water, and management to maximize the benefits of the irrigation process. Although modern irrigation systems are available, traditional irrigation is still commonly used with low efficiency (Shaki 2014). Proper management is essential to increase the efficiency of water usage. All of this requires effective management to increase water use efficiency. This can be achieved by applying statistical analysis and computer programs to analyze soil and water data, as well as studying the distribution of water among different irrigation systems. These actions can improve design and administrative aspects and ultimately increase irrigation efficiency. There has been significant development in software and computer technology. This section will focus on three types of management: water resource management, irrigation management, and crop water requirements.

2.5 Water Resources Management

Modelling groundwater is necessary to predict discharge from the aquifer. Psilovikos (1999) showed that combining simulation and optimization methods can effectively plan and manage groundwater systems. This strategy can also be employed in the Murzuq basin for better water resource management. The development of groundwater resources progresses through three major stages: exploration, evaluation, and exploitation or management (Freeze and Cherry 1979). During the exploration stage, geological and geophysical techniques are used to search for suitable aquifers. Surface geological methods are used initially based on available maps and reports. However, subsurface geological methods require test wells or holes, especially for large-scale well fields. These methods help to describe subsurface conditions in terms of permeability and hydraulic conductivity and provide the opportunity to carry out geological and geophysical loggings such as resistivity and gamma ray logs, which have become standard tools in groundwater exploration. Test wells can also collect core samples of geologic material and water samples for chemical analysis and indicate the elevation of the water table.

The evaluation stage involves measuring aquifer parameters such as transmissivity, storability, aquifer yield, and designing wells. The exploitation or management stage involves making decisions about water withdrawal in an economically feasible way that does not compromise the long-term viability of the groundwater reservoir. The success of the management stage depends on the accuracy of the aquifer parameters determined in the evaluation stage, so these two stages are often considered together. Therefore, the evaluation and management stages are often linked because they both involve addressing the same questions:

1. What are the best locations and pumping rates for the wells?

2. What will be the effect of the pumping rate on regional water levels?
3. What are the long-term yield capabilities of the aquifer?
4. Are there any undesirable environmental effects such as seawater intrusion?
5. Does the exploitation of the ground water have any effect on the other components of the hydrological cycle or on connected ecosystems?
6. Does the exploitation have effect on earth subsidence?

These questions can be answered by applying some techniques, such as simulation, which can predict the effect of the pumping rate on the water level (Fetter 2001). The best patterns of well locations and pumping rates can be achieved using an optimization model. However, it is necessary to explain terms related to groundwater management strategies such as safe yield and sustainability before proceeding.

3 Irrigation Management

Irrigation management refers to the process of efficiently and effectively managing the water resources used for irrigation purposes. It involves keeping track of water usage, monitoring soil moisture levels, and implementing strategies to optimize water distribution. The Land and Water Development Division of the FAO developed a decision support system model (i.e., CROPWAT model) to calculate reference Evapotranspiration, crop water requirements and crop irrigation requirements to develop irrigation schedules under various management conditions and scheme water supply and to evaluate rainfed production, drought effects and efficiency of irrigation practices. This model predicts crop yields by calculating ETa and Ks based on actual and potential yield ratios when all climate, soil, and crop parameters are available.

The model has successfully analyzed crop water requirements in Kenya (Karanja 2006), Senegal (Diop 2006), Canada (Doria et al. 2009), Syria (Ibrahim and Yacoub 2009), and China (Feng et al. 2009). The CROP WATT has become a widely employed standard for creating synthetic water production functions, which remains in use to this day. It enhances the viability of this method for broad planning, project design, and swift assessment applications, often resulting in a preliminary estimation at the first order. The Libyan General Water Authority conducted a study using weather station data from various parts of Libya to determine the water requirements and irrigation schedules of essential crops. The result of this study is now used as a reference for crop water requirements in Libya.

Over the last four decades, the use of CROPWAT has allowed for the analysis of the relationship between crop yield and water use. With technological advancements, novel needs have arisen for planning and managing water in agriculture, including those related to climate change. Therefore, FAO has developed a crop simulation model named AquaCrop to improve the quantification of crop yields in response to water use and water deficit. AquaCrop model uses the original equation of Doorenbos and Kassam (1979) as a point of departure and evolves from it by calculating the crop biomass based on the amount of water transpired and the crop yield as the proportion of biomass that goes into the harvestable parts. The model outputs have been calibrated and tested, providing confidence in its performance. This model simulates the growth, productivity, and water consumption of crops in response to varying environmental conditions and water availability. The accessibility of this model to various users and applications has improved with the development of standard crop parameters. It also enables assessment of responses to altered water and temperature regimes and elevated carbon dioxide concentrations under various climate change scenarios.

4 Application Management

As mentioned before, sprinkler and center-pivot irrigation systems have become widely used in southern Libya in recent decades. These irrigation methods have proven to be efficient in providing water to crops and effectively managing water resources. So, the assessment of these systems has gained significant importance. The evaluation in the field of operating systems is one of the most critical criteria, which is the uniformity coefficient, as confirmed by Soloman (1984). The distribution of water in sprinkler irrigation systems is affected by many factors, including the operating pressure, the distance between the sprinklers and their height above the ground, the speed of the pivot sprinkler, wind speed, and temperature (Camp and Sadler 1994).

In (2012) Shaki conducted a survey that showed irrigation causes significant water losses in Murzuq Basin. Notably, the study highlighted that the efficiency of water utilization (EWU) remains suboptimal, with an average of approximately 57% over the observed project durations. Thus, the study recommended conducting applied research on water, soil, and various crops to develop a strategy and sponsorship based on the optimal use of these scarce natural resources. To improve the long-term sustainability of Murzuq's water resources, the maintenance of pumps and conveyance facilities should be enhanced to raise the efficiency of water utilization (EWU) and the experience of farms (Shaki and Adeloye 2006). Most water loss is caused by the water distribution networks and pump and connection corrosion, necessitating regular maintenance and replacement of worn parts.

5 Conclusion

Managing scarce water supplies is essential for sustained agriculture in arid and semi-arid areas. Due to the ongoing water shortage in Libya, effective water management methods must be implemented to meet the growing water demand in the country. In order to ensure the sustainability of water resources in Libya, this chapter stresses the importance of sound water management practices. It also highlights the necessity of precise data gathering, effective water distribution networks, and cutting-edge modeling methodologies to maximize water usage in the nation. The agricultural sector in such areas may have trouble satisfying food production needs and establishing economic stability without effective water management measures. Since irrigation relies on groundwater, especially fossil water, monitoring and controlling usage is necessary to maintain long-term sustainability and prevent depletion. Implementing efficient irrigation practices and promoting water conservation techniques can significantly mitigate the effects of water scarcity in the region. Advanced irrigation systems, such as the canter-pivot and sprinkler systems, can help farmers maximize water efficiency by delivering water directly to the roots of plants and avoiding unnecessary evaporation. In addition to these technological solutions, effective water management strategies entail implementing policies and regulations that prioritize sustainable water use and address the competing demands for water resources in the country.

References

Alker M (2008) The Nubian sandstone aquifer system. In: Conceptualizing cooperation on Africa's transboundary groundwater resources. Ed: Waltina Scheumann and Elke Herrfahrdt-Pähle. German Development Institute. http://www.isn.ethz.ch/. Accessed on 21 May 2010

Camp CR, Sadler EJ (1994) Center pivot irrigation systems for site-specific water and nutrient management. St. Joseph. Mich, ASAE

CEDARE (2014) Libya water sector M&E rapid assessment report, monitoring & evaluation for water in North Africa (MEWINA) Project, Water Resources Management Program, CEDARE

Diop M (2006) Analysis of crop water use in Senegal with the CROPWAT model. Agro-climatologist, Institut Sénégalais de Recherches Agricoles (LERG/ISRA), Campus universitaire del'ESP

Doorenbos J, Kassam AH (1979) Yield response to water. FAO Irrigation and Drainage Paper, No. 33. FAO, Rome, 172 p

Doria R, Madramootoo CA, Mehdi BB (2009) Estimation of future crop water requirements for 2020 and 2050 using CROPWAT. Brace Center for Water Resources Management, Macdonald Campus of McGill University, Montreal, Canada

United Nations Environment Programme (2010). Ancient Water is Used to Irrigate a Desert—Murzuq Basin, Libya – UNEP Global Environmental Alert Service (GEAS) October 2010. https://wedocs.unep.org/20.500.11822/40790.

FAO (1997) Irrigation in the near east region in figures-Libya. Food and Agriculture Organization of the United Nations, Rome. Available at: http://www.fao.org/docrep/W4356E/w4356e0j.htm#Libya

Feng Z, Liu D, Zhang Y (2009) Water requirements for irrigation. Scheduling of spring maize using GIS and CROPWAT model in Beijing-Tianjin-Habei region. Earth Environ Sci Chinese Geographical Sci 17(1):6–63

Fetter (2001) Applied hydrogeology, 4th ed. Prentice Hall

Fleury P, Pistre S, Bakalowicz M (2023) Coastal karst aquifers and submarine springs: what future for their water resources? Comptes Rendus. Géoscience 355(S1):1–14

Ibrahim B, Yacoub A (2009) Using CROPWAT model for scheduling irrigation and determining water use of cotton in two areas of Syria. Arab Gulf J Sci Res 27(1–2):70–77

IMB Buchholz Consulting Engineers (1977) The consulting services for the Grain production project of wadi Barjuj and wadi Irawan

Karanja FK (2006) Cropwat model analysis of crop water use in six districts in Kenya. CEEPA discussion paper No. 35. ISBN 1-920160-35-3, Center for Environmental Economics and Policy in Africa, University of Pretoria in Senegal with the CROPWAT model. Agro

LIB/00/004. Atlas of Natural Resources for Agricultural use in Libya. 2009. Available online: http://www.fao.org/geospatial/resources/detail/en/c/1024739/. Accessed on 11 Dec 2018

Libya. http://www.water-technology.net/. Accessed 21 May 2010; World Bank (2010) World Bank development indicators. http://data.worldbank.org/. Accessed on 21 July 2010

Liu Y, Jiang X, Zhang G, Xu YJ, Wang X, Qi P (2016) Assessment of shallow groundwater recharge from extreme rainfalls in the Sanjiang plain, northeast China. Water 8(10):440

MEWINA Libya (2013) MEWINA Libya national task force responses to the RAR short form Questionnaire

NSIWRM (1999) National strategy for integrated water resources management (2000–2025). Libya

Shaki A (2012) Irrigation practices in southwestern Libya. Libyan Agric Res Center J Int 3(S2):1474–1482. ISSN 2219-4304; IDOSI Publications. https://doi.org/10.5829/idosi.larcji.3.S2.25 (in Arabic)

Shaki A (2014) Uniformity distribution of irrigation water as an indicator of water management in the Wadi Al–Hayat- southwest Libya. Egypt J Appl Sci 29(8):802–811p (in Arabic)

Shaki A, Abdalla A (2021) Groundwater evaluation for irrigation in Wadi al Shati. Libyan J Ecol Environ Sci Technol (LJEEST) 3(3 Dec):A-(56–60)

Shaki A, Adeloye A (2006) Evaluation of quantity and quality of irrigation water at Gadowa irrigation project in Murzuq Basin, southwest Libya. Agric Water Manag 84:193–201

Soloman KH (1984) Yield related interpretation of irrigation uniformity and efficiency measure. Irrigation Sci 5(3):161172

Zurqani HA (ed) (2021) The soils of Libya. Springer

Zurqani HA (2022) Introduction to environmental applications of remote sensing and GIS in Libya. Environmental applications of remote sensing and GIS in Libya. Springer International Publishing, Cham, pp 1–9

Zurqani HA, Mikhailova EA, Post CJ, Schlautman MA, Elhawej AR (2019) A review of Libyan soil databases for use within an ecosystem services framework. Land 8(5):82

Dr. Ali A Shaki is an associate professor of Water science in Faculty of Environment and natural resources, Wadi Al-shatti University, Wadi Al-shatti, Libya. He got a Ph.D. degree in Water Resources Management at Heriot Watt University, UK in 2006. On scope of ground water management, Appling simulation and optimization models on Murzuq basin southwest of Libya. His MSc degree from Agriculture Faculty, Tripoli University, Tripoli, Libya in 1997, on scope of Evaluation water for irrigation. Dr. Shaki is a director of postgraduate office in Faculty of Environment and Natural resources and a member of postgraduate committee in Wadi Al-shatti University, From 2022 until now. He worked an academic staff member in soil and water department at Faculty of agriculture in Sebha University, from 1997 to 2022. He was a member of consultant committee Ministry of agriculture, Livestock and Marine November 2013–October 2016. He work as a team leader in irrigation water management in south of Libya in Water Harvesting and Irrigation Management Project, Collaborative Program between Agriculture Research Canter (ARC)—Libya and/International Canter for Agricultural Research in the Dry Areas (ICARDA) from February 2009 to October 2012.

Dr. Hamdi A. Zurqani is an Assistant Professor of Geospatial Science in Natural Resource Management and Conservation at the University of Arkansas Agricultural Experiment Station, Arkansas Forest Resources Center, University of Arkansas at Monticello, USA. With a strong academic foundation and a passion for innovative technologies, he is also an FAA (i.e., Pt107) Certified sUAS/Drones Pilot, leveraging this expertise to enhance his proficiency in remote sensing, GIS, and geospatial data analysis, which he applies to advance his research and teaching in environmental monitoring and analysis. Dr. Zurqani is a recognized expert as a result of his internationally acclaimed work in the areas of environmental information science, remote sensing, geospatial analysis, land evaluation, sustainability, pedology, and soil science education. He has conducted research across the world, including the United States of America and Africa, and has served as PI, co-PI, or co-investigator on several grants-funded research projects. In addition to his research and teaching responsibilities, Dr. Zurqani is actively involved in institutional service at the University of Arkansas at Monticello. He serves on the UAM Campus-wide Artificial Intelligence (AI) Review Committee, the UAM Faculty Assembly Student Engagement Committee, and the UAM Library Committee.

Professionally, Dr. Zurqani contributes to the broader academic community through a variety of leadership and collaborative roles. He is an active member of the Steering Committee of the Institute of Electrical and Electronics Engineers (IEEE) Mediterranean and Middle-East Geoscience and Remote Sensing Symposium (M2GARSS) and serves as an Ambassador for the Geoscience and Remote Sensing Society (GRSS) Chapter of the IEEE in Libya, facilitating communication, collaboration, and knowledge-sharing within the geospatial and remote sensing communities. In recognition of his expertise and substantial contributions to the field, Dr. Zurqani was invited in 2024 to serve on the Technical Program Committee for the IEEE Asia-Pacific Conference on Geoscience, Electronics, and Remote Sensing Technology (AGERS). In 2025, he also joined the IEEE GRSS Technical Committee on Remote Sensing Environment, Analysis, and Climate Technologies (REACT), a platform for scientists and engineers to exchange ideas and knowledge on environmental issues and the impacts of climate change, emphasizing support for sustainable development goals (SDGs) through remote sensing technologies. Beyond these roles, Dr. Zurqani holds the position of UAM Co-Principal Investigator at the Southeast Climate Adaptation Science Center, where he is involved in research initiatives focused on climate adaptation and resilience in the southeastern United States and its surrounding regions.

Dr. Zurqani's editorial and peer-review activities further highlight his commitment to advancing scientific research. He serves on the Editorial Boards of several journals, including *Frontiers in Earth Science* (Geoinformatics), *Discover Soil* (Springer Nature), *Frontiers in Remote Sensing*, and *Remote Sensing* (MDPI). He has also been appointed as a Guest Editor for special issues in several journals, focusing on topics such as advancing spatial prediction of soil properties using remotely sensed data and geospatial artificial intelligence, and UAV agricultural management. Dr. Zurqani's dedication to his field is further demonstrated by his role as a peer reviewer for numerous publishers, including Springer Nature, Nature, Frontiers, Elsevier, and MDPI, among others. He is also the editor of several influential publications that address critical environmental challenges. His notable works include: *The Soils of Libya* (2021), part of Springer's World Soils Book Series; *Environmental Applications of Remote Sensing and GIS in Libya* (2022); and *Water Resources of Libya: Challenges and Management* (2025), published as part of the Springer Water Series (SPWA). These publications highlight his multidisciplinary expertise and commitment to addressing top environmental issues in Libya and beyond.

An Overview of Flood Hazards in Libya: Impacts and Required Actions

Hamdi A. Zurqani, Abdulsalam Al-Bukhari, Shadia A. Alzurqani, Mhi Eldin M. Alkhboli, and Ali A. Bataw

Abstract Floods are a common natural disaster that happens when an excess of water covers land that is usually dry. Heavy rainfall or a storm surge in coastal areas can lead to flooding, causing damage to homes, infrastructure, and the environment. These triggering events, coupled with the response of the basin, including factors such as slope and land use, which may influence the infiltration and retention capacity of the soil, may lead to flooding. Climate change is also causing more frequent and severe floods, impacting communities worldwide. This chapter provides an overview of flood hazards in Libya. It also discusses the causes, impacts, and required actions to reduce flood risk in the country. The chapter further emphasizes the importance of effective flood risk management strategies. Besides, it highlights the need for comprehensive planning, infrastructure development, early warning systems, and community engagement to mitigate the devastating effects of floods in the country.

H. A. Zurqani (✉)
University of Arkansas Agricultural Experiment Station, Arkansas Forest Resources Center, College of Forestry, Agriculture, and Natural Resources, University of Arkansas, Monticello, AR, USA
e-mail: Hzurqani@uark.edu; Zurqani@uamont.edu

S. A. Alzurqani
College of Forestry, Agriculture, and Natural Resources, University of Arkansas at Monticello, 110 University Court, Monticello, AR 71656, USA
e-mail: sz047117@uamont.edu

A. Al-Bukhari
Department of Forest and Rangeland, Faculty of Natural Recourses and Environmental Sciences, University of Omar Al-Mukhtar, Al Bayda 00218, Libya
e-mail: a.albukhari@omu.edu.ly

M. E. M. Alkhboli
Department of Soil and Water Sciences, Faculty of Agriculture, University of Omar Al-Mukhtar, Al Bayda 00218, Libya
e-mail: mhieldin.alkhboli@omu.edu.ly

A. A. Bataw
Department of Zoology, Faculty of Sciences, University of Omar Al-Mukhtar, Al-Bayda 00218, Libya
e-mail: ali.bataw@omu.edu.ly

H. A. Zurqani (ed.), *Water Resources of Libya*, Springer Water,
https://doi.org/10.1007/978-3-031-80920-0_8

Keywords Water resources · Water scarcity · Climate change · Mediterranean Sea · Sea level rise · Flash flood · Flood preparedness and response

1 Introduction

Floods are a natural hazard that can devastate human health, infrastructure, and the environment. Flooding can be caused by a variety of factors, including prolonged heavy rainfall, highly accelerated snowmelt, severe winds over water, unusually high tides, tsunamis, or the failure of dams, levees, retention ponds, or other water-retaining structures (Hirschboeck 1991). While floods are common in tropical humid regions, arid and dry climates are not exempt from flood occurrence and effects (Lin 1999). For example, Africa is known for its dry, arid weather, and drought is the mainly known and studied natural hazard. Most studies and research only address this problem with little focus on floods when many African regions have experienced devastating flood events. The World Meteorological Organization (WMO) 2023 report states that Africa is highly vulnerable to natural disasters, particularly floods and droughts. There has been a comment on the increase in flood fatalities in several African countries from 1990 to 2009 (Tramblay et al. 2020). In many African countries, floods pose a significant natural hazard (Parker and Thompson 2000). Vousdoukas et al. (2022a, b) identify the Nile River basin, coastal areas of West and East Africa, and the Sahara Desert borders as regions prone to flooding.

Libya is one of the African countries that has experienced flood events throughout its history, highlighting its vulnerability to these extreme weather events (Harmon 2014). Libya is characterized by its vast arid landscapes and coastal plains that face a growing challenge in the form of flood hazards. While Libya is often associated with its dry climate, the country is not immune to periodic and frequently devastating floods, and it has experienced several significant floods in recent years (Oduoye et al. 2024). Floods in Libya have historically been sporadic but can be severe when they occur. The impacts of such events extend beyond immediate damage to property and infrastructure, often causing displacement, loss of livelihoods, and an increased burden on already limited resources. In addition, the changing climate patterns in the region have the potential to exacerbate the frequency and intensity of flooding, making it imperative to address these hazards proactively. The areas that are most prone to flooding include coastal areas, dry riverbeds (wadis), and highland areas where heavy rainfall is common (Zeleriakoval et al. 2012). Particularly, the eastern region of Al Jabal Al Akhdar highland has had the highest recorded annual precipitation, followed by the western region of Jabal Nafusa highland (Abushnaf 2014).

The floods can cause widespread damage and displacement and impact cultural heritage, human mobility, and the lives and livelihoods of local communities (House 2007). In addition to the immediate impacts, floods can also have long-term consequences. For example, floods can contaminate water supplies, leading to disease outbreaks. It can also disrupt food supply chains and damage agricultural land, leading to food insecurity among vulnerable populations (Alderman et al. 2012). In Libya, the lack of available data makes it challenging to accurately estimate the frequency of floods and their impact on the country. Hence, this chapter aims to shed light on the challenges posed by flood hazards in Libya and provide an overview of flood hazards in the country. It also discusses the causes and impacts of floods and the required actions to reduce flood risk in the country.

2 Causes of Flooding in Libya

Floods can occur in Libya in both the coastal and desert regions. The most flood-prone areas in Libya are the coastal plains, the Jufra Depression, and the Fezzan region. Heavy rainfall and storms can result in floods in the coastal areas when Wadis (dry riverbeds) overflow due to excessive rain in different parts of the country. Such weather conditions can result in flash floods that carry debris and sediment from the wadi floor along with the water (Hsiao et al. 2021). In addition, outdated and insufficient drainage systems exacerbate the problem. In desert regions, flash floods are common due to sudden and intense rainfall that the arid soil cannot absorb. These floods can be dangerous as they rapidly fill Wadis and cause water to rush through narrow canyons, posing a threat to nearby communities (Zurqani et al. 2022). In mountainous regions, the combination of steep slopes and heavy precipitation can result in devastating landslides. The loose soil and rocks can easily be dislodged, burying homes and blocking roads (Svalova et al. 2019). Overall, the country's diverse geography presents a constant challenge to managing and mitigating the risks associated with natural disasters (Zurqani et al. 2019). Flooding in Libya can be caused by several factors, including:

2.1 Heavy Rain

Despite being a primarily desert country, Libya still receives some rainfall that can result in flooding in areas with inadequate drainage systems. These heavy rainfalls are commonly caused by powerful storms, referred to as "Mediterranean cyclones." They form over the warm Mediterranean Sea and travel eastward, bringing heavy rain and strong winds to the Libyan coast (Nastos et al. 2018). While infrequent, these storms can be responsible for some of the heaviest rainfall events in the region. In addition, the proximity to the Mediterranean Sea exposes Libya to the possibility of coastal flooding during storms and high tides, particularly in low-lying areas (Vousdoukas et al. 2022a, b).

The majority of rainfall in Libya occurs during the winter months (October to March), with a significant decrease in precipitation during the summer months (June to August (Zurqani 2021). Rainfall distribution across Libya is uneven (Zeleriakoval et al. 2012). The coastal regions receive slightly more precipitation compared to the interior due to the influence of the Mediterranean Sea and localized weather patterns in mountainous areas (Saad et al. 2011). These heavy rain events can cause flash floods to develop within minutes or hours in the mentioned regions, depending on the intensity and duration of the rain. Furthermore, flash floods can occur in wadis (dry riverbeds) during sudden downpours, leading to a rapid rise in water levels. This poses a significant risk to nearby communities and infrastructure.

The World Bank (2023) found that floods are Libya's most frequent natural disaster, affecting an average of 10,000 people annually. The study also found that floods have caused an average of $100 million in damage annually. The impacts of floods in Libya can be far-reaching. In 2018, for example, a flood in the city of Benghazi killed at least 13 people and displaced thousands more. The flood also caused extensive damage to infrastructure, including roads, bridges, and homes. Furthermore, a study by Zamot and Afkareen (2020) documented a flood that hit the east region of Al Jabal Al Akhdar, where a devastating flood occurred on September 27th, 2018, in Mikhili village. The flood caused significant damage to properties and lives and washed away cars belonging to both travelers and residents. Flash floods in the east are a well-known phenomenon, and they cause much damage every year.

In December 2013, Heavy rains caused flash floods in southwestern Libya, particularly around the city of Sabha, which is known as an arid area. Also, Tripoli, the capital city, experienced flash floods and heavy rainfall, marking some of the heaviest seen in Tripoli city in 40 years (FloodList 2013). While not as widespread or devastating as the recent floods, these events resulted in the deaths of at least 16 people, displaced many others in Sabha, and caused infrastructure damage in Tripoli. The most recent flood in the country was in 2023 as a consequence of Daniel Storm that struck the east region of Libya, with winds of 70–80 km/h and up to 240 mm rainfall, causing large-scale destruction over several regens as Benghazi, Al-Marj, Al-Bayada, Takenis, Al-Wardiya, Al-Bayda, Shahat, Sousse, Al-Makhili, Derna. The affected areas are the home of about 1.5 million people. According to the statistics of the Libyan Red Crescent, Al-Bayda Branch (2024), most areas of Al-Jabal Al-Akhdar were damaged by floods in varying proportions, ranging from 10 to 90%, causing a number of deaths and missing persons, as shown in the Table 1 in addition to damage to property and infrastructure.

Table 1 Summary statistics of affected areas in Al-Jabal Al-Akhdar by floods according to the Libyan Red Crescent, Al-Bayda Branch (10–14 September 2023)

Regions	Damage rate (%)	Affected families	Missing	Deaths
Al-Bayada,	40	600	37	32
Aslanta	20	25	2	2
Gundula	80	175	3	9
Marawa	10	10	0	1
Omar Mukhta	50	80	17	3
Al-Wardiya	90	75	15	7

The coastal city of Derna was the most affected, as its dams were severely damaged. It resulted in a significant humanitarian crisis. About 4352 people lost their lives, and more than 43,400 people have been internally displaced. About 884,000 people live in five counties in areas affected to varying degrees (OCHA and WHO 2023). The disaster also caused widespread destruction to homes and infrastructure, leaving many displaced and impacting essential services like healthcare and sanitation. More than 18,500 houses are estimated to have been destroyed or damaged (IOM 2023). The World Bank, the UN, and the European Union (2023) reported damage to infrastructure sectors, including energy and transport, and damages to assets in the agriculture, health, and education sectors. The cost of the assets of Libya's electricity infrastructure damage is about $34 million, including power transmission facilities, and the total road network damage entirely or partially in the affected regions is estimated at 673 km. Also, around 200 educational facilities were entirely or partially destroyed in 20 municipalities. Figure 1a–d illustrates examples of extensive damage to infrastructure, including bridges, homes, and roads over the affected areas.

(a) Bridges located in the eastern part of the city of Sousse, Libya (Photo courtesy: Younis Ahweij).

(b) Homes located in the Al-Gharika neighborhood of Al-Bayda, Libya (Photo courtesy: Libyan Red Crescent, Al-Bayda Branch).

(c) Roads in Al-Gharika neighborhood, Al-Bayda, Libya (Photo courtesy: Libyan Red Crescent, Al-Bayda Branch).

(d) Al-Bayda Western Entrance (i.e., City center), Al-Bayda, Libya (Photo courtesy: Libyan Red Crescent, Al-Bayda Branch).

Fig. 1 Examples of extensive damage to infrastructure, including bridges, homes, and roads over the affected areas (10–14 September 2023)

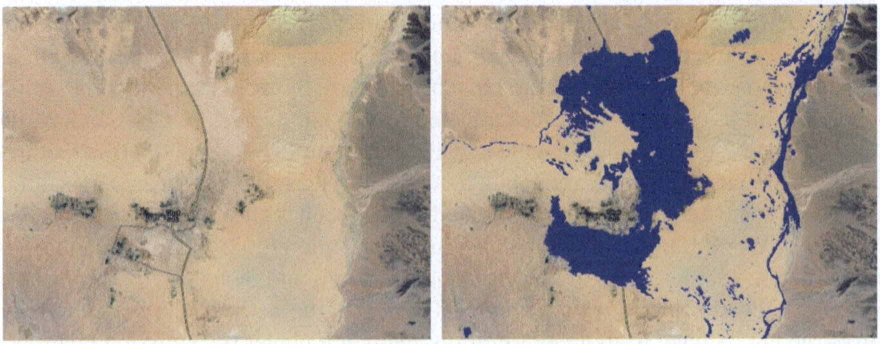

(a) Pre-event (25 May 2019) **(b)** Post-event (7 June 2019)

Fig. 2 The area of the city of Ghat, southwestern Libya, **a** pre-event, and **b** post-event

In 2022, Zurqani et al. conducted a study using Google Earth Engine and deep learning techniques to analyze and map the extent and impact of the 2019 flood disaster in the southwestern Libyan city of Ghat. The study revealed that the flood extent extraction analysis results indicated that approximately 2255.67 ha of the study area were flooded during the wave of heavy rainfall and thunderstorm events in June 2019, causing flooding and damage in several locations around the city (Fig. 2a, b). During this event, 70% of urban areas and roads were affected by floods, followed by half of the shrubs area inundated. About 30% of agriculture, tree canopy, and barren land were flooded overall land cover classes, while the sand dunes had less area affected.

2.2 Dam Failures

Libya meets its water needs from four sources: groundwater (95%), surface water (rainwater and dams), desalinated seawater, and wastewater recycling (Abdudayem et al. 2014). Libya has a few dams that are intended to be used to store water for irrigation and drinking water supply. Water collected in these major dams is used for agricultural water supply, industrial projects, and domestic use. Approximately 16 significant dams have been built in Libya to collect rainwater, with a total capacity of 385 million m^3 and an average yearly storage capacity of 61 million m^3. The largest three dams are Wadi Quattara, Wadi Kaam, and Wadi El-Magineen, having a design capacity of 135, 111, and 58 million m^3, respectively (Brika 2018). However, these dams are often poorly maintained and can fail during heavy rainfall, leading to devastating floods downstream. The two earth fill dams on the Wadi Quattara were built between 1969 and 1972, but neither was filled. The secondary Dam failed in December 1977 following a partial filling below average water level (Binquet and Zappi 1988). This highlights the impact of flooding being worsened by outdated

(a) Abu Mansour dam, Derna, Libya (Photo courtesy: Younis Ahweij, 25 October 2023).

(b) Al-Bilad dam, Derna, Libya (Photo courtesy: Younis Ahweij, 25 October 2023).

Fig. 3 The collapse of the dam Dams in Derna: **a** Abu Mansour dam and **b** Al-Bilad Dam

infrastructure, such as inadequate drainage systems and poorly constructed buildings in the country. Climate change is increasing the risk of flooding due to the frequency and intensity of rainfall events in the region. As shown in Fig. 3a and 2b, the collapse of Abu Mansour and Al-Bilad Dams in Derna after the Daniel Storm in 2023 illustrates the devastating consequences of dam failures (Ashoor and Eladawy 2024).

This disaster resulted in the displacement of thousands of people and significant infrastructure damage. Poor maintenance practices and lack of investment in dam infrastructure can lead to structural failures, highlighting the urgent need for improved dam maintenance across the country. In addition, implementing effective flood management strategies, such as enhancing natural Wadi (dry riverbeds) courses and building flood-resistant structures, can help mitigate the devastating impacts of flooding in Libya.

Derna is one of the Libyan cities that has seen several flooding incidents caused by dam collapses. Some of these incidents were very devastating, resulting in severe damage to infrastructure and loss of human life. The Wadi Derna, which originates in the Jebel Akhdar mountains and flows a distance of about 75 km (47 miles) before emptying into the Mediterranean Sea near the port city of Derna, poses a well-known risk of flooding. The Wadi had big floods in 1968 and 1986. The city has historically seen occasional flooding, especially during the winter months. During World War II, when German forces controlled the city, it had a catastrophic flood in 1941. There were no reports of any human casualties.

According to Libyan historian Faraj Daoud al-Darnawi, another major flooding occurred in Derna in 1959, killing and injuring hundreds of people (Anadolu agency, 2023). In response to the frequent floods experienced by the city, two dams were constructed in the 1970s, the Al-Bilad Dam and the Abu Mansour Dam. These dams were built to regulate water flow during heavy rainfall, thus reducing the risk of devastating floods and protecting the city from potential damage. Also, to manage water resources and address water scarcity in the region. The flood was caused by heavy rain produced by Storm Daniel in September 2023. It was the most devastating flood the city experienced, which was a powerful Mediterranean storm. The storm dropped an immense amount of rain, exceeding eight months of average rainfall in just 24 h (Imran et al. 2024). This excessive rainfall overwhelmed the region's infrastructure, causing widespread flooding and the collapse of the two dams near Derna. The flooding resulted in a significant humanitarian crisis, according to the United Nations. The flood also displaced tens of thousands of people, who were forced to flee their homes and seek shelter in temporary camps. The 2023 Derna flood was the worst natural disaster to hit Libya in decades. The flood caused widespread damage and loss of life, and it will take years for the city to recover.

2.3 *Climate Change*

Climate change is expected to increase the frequency and intensity of extreme weather events, including floods (Archer et al. 2024; Smith et al. 2024; Zhou et al. 2023). The increase in temperature and changing rainfall patterns can lead to more intense rainstorms and flash floods in the flood-prone areas of Libya. Furthermore, the sea level rise caused by climate change can exacerbate coastal flooding and threaten low-lying regions (Vousdoukas et al. 2022a, b). Libya is located along the coast and is facing a dangerous threat from the rise of sea level. As the sea level increases, tides and storm surges increase, leading to more frequent and severe flooding. Besides, sea level rise can negatively impact the environment, including saltwater intrusion into freshwater aquifers (White and Kaplan 2017). This can contaminate drinking water sources and agricultural lands, worsening the effects of flooding. In addition, coastal erosion due to rising sea levels can weaken infrastructure and further exacerbate the risk of flooding.

Moreover, according to the United States National Oceanic and Atmospheric Administration (Sweet et al. 2017), the rise in air and water temperatures will contribute to a global increase in sea levels, which could reach 0.5 m by 2050. This is a major concern for Libya as sea-level rise will cause erosion of shorelines, flood low-lying coastal areas, and impede the functioning of coastal infrastructure. A recent study conducted by Zurqani et al. in 2019 revealed that the rise in sea level caused by global climate change puts many coastal Libyan cities at risk. The cities at the highest risk are Zuwarah, Misurata, Ras Lanuf, Aqilah, Brega, and Zaytiniyya (Fig. 4). The study predicts that with a rise in sea level of 1, 3, and 6 m, the affected areas could lose 700 km^2, 3200 km^2, and 7200 km^2 of land, respectively. These cities are not only significant in terms of their cultural and economic value but also require urgent attention to develop effective coastal management strategies. According to the study, the rise in sea level poses significant risks to the infrastructure, human settlements, and natural ecosystems of coastal cities. These risks include an increase in flooding, erosion of coastal lands, and intrusion of saltwater into freshwater sources.

Without adequate measures in place to manage these risks, the already challenging task of mitigating natural disasters in Libya will become even more difficult. Therefore, it is essential for the government of Libya to implement coastal protection measures, such as building seawalls and reinforcing coastal structures, to mitigate the impact of sea level rise and reduce the vulnerability of coastal communities to

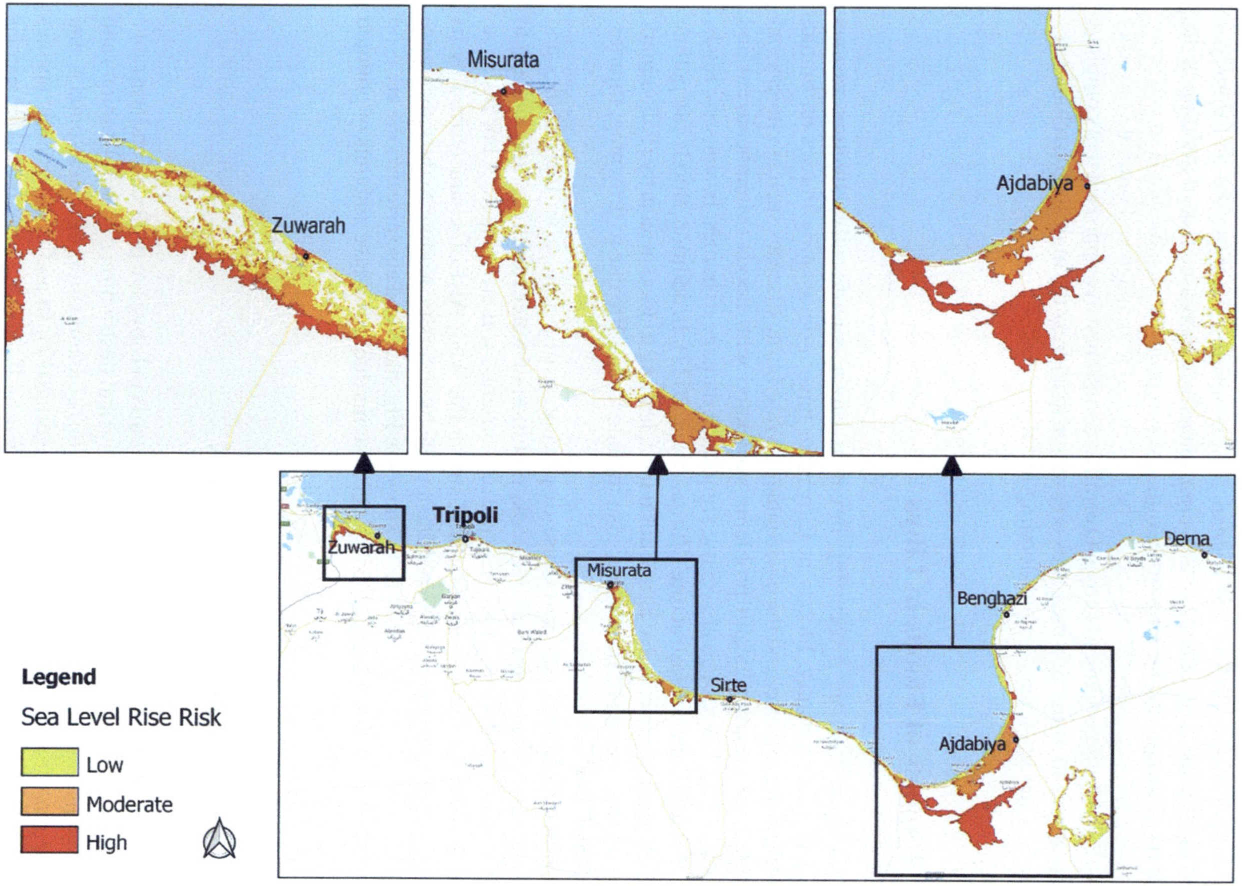

Fig. 4 Projections of the future sea rise due to climate change along the Mediterranean Sea coast of Libya: **a** 1 m sea rise (affected area: 700 km^2), **b** 3 m sea rise (affected area: 3200 km^2), **c** 6 m sea rise (affected area: 7200 km^2) (adapted from: Zurqani et al. 2019)

flooding (Dedekorkut-Howes et al. 2020). In addition, investing in improved water management systems and infrastructure can help mitigate the risks of saltwater intrusion and ensure a sustainable supply of freshwater for both human consumption and agricultural needs (White and Kaplan 2017). Besides, deforestation and improper land management practices can also contribute to an increased risk of flooding in the country (Zurqani 2021). Therefore, initiatives to promote sustainable land use practices and reforestation efforts should be prioritized to manage and mitigate the risks associated with flooding in Libya. Overall, a comprehensive approach that combines adaptation and mitigation measures is essential to address the multifaceted challenges posed by the sea level rise and flooding in Libya.

3 Impacts of Floods in Libya

Floods have become a recurring threat in Libya, significantly affecting various aspects of the country. Floods in Libya can have a devastating impact on human health, infrastructure, and the environment. Waterborne diseases such as cholera and typhoid are common health problems associated with floods (Mtembenuzeni and Kushe 2018). Floods can also cause significant damage to homes, businesses, roads, bridges, and other infrastructure (Zurqani et al. 2022). In addition, floods can pollute water supplies and damage agricultural land, leading to food insecurity (Falkenmark 2001). The impacts of floods may also include loss of life due to drowning, injuries, and other health problems.

Moreover, floods can also cause displacement of people from their homes and businesses, loss of crops and livestock, pollution of water resources, and damage to ecosystems (biodiversity loss, loss of soil nutrients, soil erosion, etc.). For instance, floods swept away the vegetation and soil in the Wadi Al-Kof, characterized by its dense vegetation cover and incredible unique biodiversity, leading to the appearance of a layer of rocks in some parts, as shown in Fig. 5a, b and also affected the areas on the coastal area. Such as habitats and ecosystems due to massive amounts of polluted sediments and debris.

3.1 Environmental Consequences

The environmental implications of flood hazards in Libya are profound and warrant a comprehensive examination. Beyond the physical destruction of natural habitats and ecosystems, flooding can lead to soil erosion, contamination of water sources, and disruption of ecological balance (Appannagari 2017). In addition, endemic species existing in limited numbers may be impacted by floods, as numerous plant species are incapable of enduring prolonged submersion in water, thus depriving their roots of oxygen and hindering their growth and reproduction cycles, potentially resulting in the extinction of various floral and faunal species (Thomas et al. 2004). Moreover,

(a) Wadi Al-Kof, Jabal Al Akhdar, Libya (Photo courtesy: Younis Ahweij, 11 January 2024).

(b) The coastal area of Al Akhdar, Libya (Photo courtesy: Younis Ahweij, 11 January 2024).

Fig. 5 The flooding impacted Wadi Al-Kof and the coastal area

habitat fragmentation may influence ecosystems, posing a risk to species populations, especially arthropods, through divergence and possibly being eradicated by strong currents, thereby causing instability among species and establishing a new ecosystem (McLaughlin et al. 2013). Furthermore, habitat fragmentation often proves harmful to individual species due to local extinctions resulting from demographic variability, with a decreased potential for patches to be re-colonized (Harrison and Taylor 1997).

Floods have a significant impact on the biodiversity of terrestrial animals in submerged areas (Bodmer et al. 2018). The overall effect varies depending on the severity and duration of the flooding, as well as the adaptability of the local plant and arthropod species. Some species may endure temporary immersion, while others may face prolonged disturbances to their populations and ecosystems (Woodman 2015; Cavallaro et al. 2017). Furthermore, surges can devastate entire biological systems by annihilating or pulverizing the shared habitats of plants and animals, forcing them to relocate or resulting in the loss of vulnerable species; additionally, large or slow-moving arthropods are at risk of suffocation if they fail to locate higher ground to evade the rising waters (Hildebrandt 1997). Moreover, floodwaters may become contaminated with chemicals, sewage, and debris, making the water unsuitable or hazardous for plant and animal species. Flooding has the potential to disrupt the natural food chains, as marine plants and prey organisms may be carried away, leaving certain species without their typical sources of sustenance (Wootton et al. 1996). Therefore, understanding the environmental repercussions of flooding is essential for developing strategies aimed at long-term environmental restoration and conservation efforts.

3.2 Economic and Social Impacts

The complex and multifaceted impacts of flooding on the economy and society in Libya are complex and varied. Flooding not only leads to immediate and visible damages, such as the destruction of infrastructure and loss of lives, but also has long-term repercussions on the economic and social fabric of the affected areas (Hallegatte et al. 2010). The disruption of agricultural activities due to the inundation of farmlands and damage to irrigation systems not only results in immediate loss of livelihoods but also has lingering effects on food security and local economies (FAO 2021). Furthermore, the strain on emergency response systems and healthcare facilities during and after flooding events can have lasting implications for public health and well-being.

4 Proposed Strategies for Effective Flood Hazard Management

In light of the escalating flood hazards in Libya, it is imperative to develop and implement comprehensive strategies for effective flood hazard management. This section proposes various strategies that encompass both mitigation and adaptation measures. Addressing the root causes of flood hazards and enhancing the resilience of communities and critical infrastructure can reduce the devastating impact of these events.

4.1 Mitigation Strategies

Mitigating flood hazards requires a multi-faceted approach that focuses on preventing and minimizing the occurrence and impact of flooding. One key strategy is the restoration and preservation of natural floodplains and wetlands, which act as vital buffers against floodwaters and help absorb excess water. In addition, implementing proper land use planning and zoning regulations can mitigate the risk of flood damage to infrastructure and communities. Furthermore, constructing well-designed water retention and drainage systems can effectively reduce the severity of flooding in urban and rural areas.

4.2 Adaptation Strategies

Adaptation measures are essential for building the resilience of communities and ecosystems to withstand and recover from flood events. Enhancing early warning systems and developing community-based disaster preparedness plans are critical components of adaptation strategies. Furthermore, incorporating climate-resilient practices in agricultural and land management can help minimize the impact of floods on food security and livelihoods.

4.3 Identifying Vulnerable Populations and Areas

For a more comprehensive understanding of the impact of flood hazards, it is essential to identify and analyze the vulnerabilities of specific population groups and geographic regions. Factors such as access to resources, socio-economic status, and infrastructural resilience play a positive role in determining the extent of the impact on different communities. A deeper analysis of the vulnerabilities will facilitate the development of targeted and effective interventions to support the most at-risk populations.

4.4 Strengthening Institutional Capacity and Governance

An in-depth evaluation of the existing policies and institutional frameworks for flood hazard management is essential for identifying gaps and opportunities for improvement. This includes an assessment of the coordination mechanisms, allocation of resources, and integration of risk-informed approaches into governance structures. Strengthening institutional capacity is vital for ensuring the efficient and effective implementation of flood hazard management strategies.

4.5 Enhancing Community Engagement and Participation

Greater emphasis on community engagement and participatory approaches can enhance the effectiveness of flood hazard management initiatives. Involving local communities in the design and implementation of strategies fosters ownership and ensures that interventions are tailored to the specific needs and realities of the affected areas. This approach also contributes to building community resilience and promoting sustainable hazard management practices at the grassroots level.

These in-depth analyses and considerations pertaining to the economic, social, and environmental impacts of flood hazards in Libya will provide a comprehensive foundation for formulating targeted and effective strategies for mitigation and adaptation. By delving into the intricacies of the challenges posed by flood hazards, it is possible to develop holistic and sustainable solutions that address the multifaceted dimensions of this critical issue.

5 Required Actions to Reduce Flood Risk

The required actions to address flood hazards in Libya encompass a multi-faceted approach. In addition to improving early warning systems and evacuation procedures, there is a pressing need to invest in resilient infrastructure and sustainable land use planning. Furthermore, promoting community-based adaptation measures and enhancing capacity building at local levels are vital components of a comprehensive flood management strategy. There are several actions that can be taken to reduce flood risk in Libya, such as:

- **Improved early warning systems**: Early warning systems can help people to evacuate before a flood occurs.
- **Improved weather forecasts**: Weather forecasts are essential for predicting floods and are produced using models, meteorological radars, weather stations, and satellite imagery.
- **Better floodplain management**: Floodplains should be managed in a way that reduces the risk of flooding. This can include zoning restrictions, floodproofing of buildings, and the creation of green infrastructure.
- **Dam safety improvements**: Dams should be regularly inspected and maintained to ensure that they are safe.
- **Raise awareness of flood risks**: People need to be educated about the risks of flooding and how to protect themselves.
- **Adaptation to climate change**: Libya should adapt to the impacts of climate change, such as sea level rise and more intense rainfall. This can include building seawalls and other coastal defenses and developing drought-resistant crops.

6 Conclusion

Floods are the most frequently occurring and geographically widespread natural disasters that are caused by weather-related factors. Flash floods are the most dangerous kind of floods because they combine the destructive power of a flood with incredible speed. These natural disasters can have devastating effects on communities, causing damage to infrastructure, homes, and the environment. Governments and communities must invest in flood management strategies and infrastructure to mitigate the impact of future floods. In Libya, floods are a severe threat, and the country needs to take action to reduce flood risk. The actions outlined in this chapter can help to protect people, property, and the environment from the impacts of flooding. This chapter aims to contribute to the ongoing discourse on disaster risk reduction and climate resilience in the region by examining the specific challenges and opportunities for addressing flood hazards in Libya. It seeks to underscore the importance of proactive measures and collaborative efforts among stakeholders to safeguard lives and livelihoods in the face of increasing flood risks in the country.

References

Abdudayem A, Scott AH (2014) Water infrastructure in Libya and the water situation in agriculture in the Jefara region of Libya. Afr J Econ Sustain Develop 3(1):33–64

Abushnaf F (2014) A land evaluation model for irrigated crops using multi-criteria analysis. Sheffield Hallam University (United Kingdom)

Africa N (2014) Natural disasters in the Middle East and North Africa

Aldali Y, Celik A (2018) Validation of thermal models for polycrystalline photovoltaic module under Derna city climate conditions. Solar Energy Sustain Develop J 7(2):27–40

Alderman K, Turner LR, Tong S (2012) Floods and human health: a systematic review. Environ Int 47:37–47

Anadolu Agency (2023) Storm daniel leaves behind mass destruction in Libya's Derna. Retrieved December 15, 2024, from https://www.aa.com.tr/en/africa/storm-daniel-leaves-behind-mass-destruction-in-libya-s-derna/2990503

Appannagari RR (2017) Environmental pollution causes and consequences: a study. North Asian Int Res J Soc Sci Humanities 3(8):151–161

Archer L, Hatchard S, Devitt L, Neal JC, Coxon G, Bates PD, Kendon EJ, Savage J (2024) Future change in urban flooding using new convection-permitting climate projections. Water Resour Res 60(1):e2023WR035533

Ashoor A, Eladawy A (2024) Watch and upgrade or deconstruct and relocate: Derna Catastrophe lessons amid the climate-change era of unpredictable flash floods

Binquet J, Zappi U (1988) Wadi Qattarah dams case history

Bodmer R, Mayor P, Antunez M, Chota K, Fang T, Puertas P, Pittet M, Kirkland M, Walkey M, Rios C, Perez-Pena P, Henderson P, Bodmer W, Bicerra A, Zegarra J, Docherty E (2018) Major shifts in Amazon wildlife populations from recent intensification of floods and drought. Conserv Biol 32(2):333–344. https://doi.org/10.1111/cobi.12993

Brika B (2018) Water resources and desalination in Libya: a review. In: Proceedings (vol 2, No 11). MDPI, p 586

Cavallaro MC, Barnhart MC, Hoback WW (2017) Causes of rapid carrion beetle (Coleoptera: Silphidae) death in flooded pitfall traps, response to soil flooding, immersion tolerance, and swimming behavior. Environ Entomol 46(2):362–368. https://doi.org/10.1093/ee/nvw165

Climate Change and Extreme Weather (2023) World Meteorological Organization. https://wmo.int/about-us/world-meteorological-day/wmd-2022/climate-change-and-extreme-weather

Climate Knowledge Portal (2023) Libya—vulnerability. Climate Change Knowledge Portal. https://climateknowledgeportal.worldbank.org/country/libya/vulnerability

Dedekorkut-Howes A, Torabi E, Howes M (2020) When the tide gets high: a review of adaptive responses to sea level rise and coastal flooding. J Environ Planning Manage 63(12):2102–2143

Derna LY (n.d.) Climate Zone, Monthly Weather Averages and Historical Data. Weatherandclimate.com. Retrieved February 26, 2024, from https://weatherandclimate.com/libya/derna

Falkenmark M (2001) The greatest water problem: the inability to link environmental security, water security and food security. Int J Water Resour Dev 17(4):539–554

FloodList (2013) 16 Dead in Libya floods, funded by European Centre for Medium-Range Weather Forecasts. https://floodlist.com/africa/16-dead-libya-floods

Floods in Libya's Derna: Worst disaster in 21st century (n.d.) Www.aa.com.tr. https://www.aa.com.tr/en/environment/floods-in-libya-s-derna-worst-disaster-in-21st-century/2992617

Hallegatte S, Ranger N, Bhattacharya S, Bachu M, Priya S, Dhore K, Rafique F, Mathur P, Naville N, Henriet F, Patwardhan A (2010) Flood risks, climate change impacts and adaptation benefits in Mumbai: an initial assessment of socio-economic consequences of present and climate change induced flood risks and of possible adaptation options. https://doi.org/10.1787/19970900

Harmon D (2014) Libya. Simon and Schuster

Harrison S, Taylor AD (1997) Empirical evidence for metapopulation dynamics. Metapopulation Biol 27–42. https://doi.org/10.1016/B978-012323445-2/50004-3

Hildebrandt J (1997) Wie sind terrestrische Wirbellose an Überflutung angepaßt? Arbeitsberichte Landschaftsökologie Münster 18:15–25

Hirschboeck KK (1991) Climate and floods. US Geol Surv Water-Supply Pap 2375:67–88. Global Health 7(1):e0401

House QE (2007) Environmentally displaced people: understanding the linkages between environmental change, livelihoods and forced migration. University of Oxford, Refugee Studies Centre

Hsiao SC, Chiang WS, Jang JH, Wu HL, Lu WS, Chen WB, Wu YT (2021) Flood risk influenced by the compound effect of storm surge and rainfall under climate change for low-lying coastal areas. Sci Total Environ 764:144439

Imran JASIM, Nemnem A, Tanim A, Khan MS, Nahian A, Goharian E (2024) Rising waters, falling dams: deciphering the Derna flood disaster

International Decade for Action on Water for Sustainable Development, 2018–2028 (2018) Un.org. https://www.un.org/en/events/waterdecade/

IOM (International Organization for Migration) (2023) Resilience, recovery & reconstruction plan, Storm Daniel (September 2023–September 2024), IOM Libya

Libyan Red Crescent Al-Bayda Branch (2023) Statistics of areas affected by torrents and floods in the Green Mountain areas, Libyan Red Crescent, Libya

Lin X (1999) Flash floods in arid and semi-arid zones. Technical documents in hydrology

McLaughlin ÓB, Emmerson MC, O'Gorman EJ (2013) Habitat isolation reduces the temporal stability of Island ecosystems in the face of flood disturbance. Adv Ecol Res. https://doi.org/10.1016/b978-0-12-417199-2.00004-5

Mtembenuzeni J, Kushe J (2018) Theory and practice of vulnerability to disasters in flood prone areas in Chikwawa, Malawi. Environ Ecol Res 6(3):137–146

Nastos PT, Papadimou KK, Matsangouras IT (2018) Mediterranean tropical-like cyclones: impacts and composite daily means and anomalies of synoptic patterns. Atmos Res 208:156–166

National Weather Service (n.d.) Flood related hazards. National Oceanic and Atmospheric Administration. https://www.weather.gov/safety/flood-hazards

OCHA (Office for the Coordination of Humanitarian Affairs) (2023) Libya, MENA region|Storm Daniel—revised emergency appeal no: MDRLY005 (Revision #1), conflict and environment observatory, UN

Oduoye MO, Karim KA, Kareem MO, Shehu A, Oyeleke UA, Zafar H et al (2024) Flooding in Libya amid an economic crisis: what went wrong?. IJS

Parker DJ, Thompson PM (2000) Floods in Africa. Floods 1:188

Saad AMA, Shariff NM, Gairola S (2011) Nature and causes of land degradation and desertification in Libya: need for sustainable land management. Afr J Biotech 10(63):13680–13687

Smith MD, Wilkins KD, Holdrege MC, Wilfahrt P, Collins SL, Knapp AK, Sala OE, Dukes JS, Phillips RP, Yahdjian L, Gherardi LA (2024) Extreme drought impacts have been underestimated in grasslands and shrublands globally. Proc Natl Acad Sci 121(4):e2309881120

South West Libya—FloodList. (n.d.). Floodlist.com. Retrieved February 26, 2024, from https://floodlist.com/africa/south-west-libya

Svalova VB, Zaalishvili VB, Ganapathy GP, Nikolaev AV, Melkov DA (2019) Landslide risk in mountain areas. Geol South Russia 9(2):109–126

Sweet WV, Kopp RE, Weaver CP, Obeysekera J, Horton RM, Thieler ER, Zervas C (2017) Global and regional sea level rise scenarios for the United States (No. CO-OPS 083)

The Food and Agriculture Organization of the United Nations (FAO) (2021) The impact of disasters and crises on agriculture and food security: 2021. Rome. https://doi.org/10.4060/cb3673en

Thomas CD, Cameron A, Green RE, Bakkenes M, Beaumont LJ, Collingham YC et al (2004) Extinction risk from climate change. Nature 427(6970):145–214

Tramblay Y, Villarini G, Zhang W (2020) Observed changes in flood hazard in Africa. Environ Res Lett 15(10):1040b5

UN launches flash appeal to aid 250,000 Libyans hit by devastating floods|UN News (2023) News.un.org. https://news.un.org/en/story/2023/09/1140767

Vousdoukas MI, Clarke J, Ranasinghe R, Reimann L, Khalaf N, Duong TM et al (2022a) African heritage sites threatened as sea-level rise accelerates. Nat Clim Chang 12(3):256–262
Vousdoukas MI, Clarke J, Ranasinghe R, Reimann L, Khalaf N, Duong TM, Ouweneel B, Sabour S, Iles CE, Trisos CH, Feyen L (2022b) African heritage sites threatened as sea-level rise accelerates. Nat Clim Chang 12(3):256–262
White E, Kaplan D (2017) Restore or retreat? Saltwater intrusion and water management in coastal wetlands. Ecosyst Health Sustain 3(1):e01258
WHO (2023) WHO Libya Situation Update 14-September-2023.WHO country Office in Libya, WHO
Woodman JD (2015) Surviving a flood: effects of inundation period, temperature and embryonic development stage in locust eggs. Bull Entomol Res 105(4):1–7. https://doi.org/10.1017/S0007485315000243
Wootton JT, Parker MS, Power ME (1996) Effects of disturbance on river food webs. Science, New Series 273(5281):1558–1561
World Bank (2023) Libya storm and flooding 2023: rapid damage and needs assessment (English). World Bank Group, Washington, DC. http://documents.worldbank.org/curated/en/099353101242428521/IDU153d4e1711e33e145321b8881cf996ea3acf7
World Bank, the UN, and the European Union (2023) Assesses damages caused by catastrophic flooding in Libya. World Bank Group, Washington, DC. https://www.worldbank.org/en/news/press-release/2024/01/24/joint-world-bank-eu-un-report-assesses-damages-caused-by-catastrophic-flooding-in-libya
Zamot J, Afkareen M (2020) Geomorphological parameters by remote sensing and GIS techniques (A case study of flash flood in Mikhili Village, Al Jabal Al Akhdar, NE of Libya). In: The fourth international conference for geospatial technologies–Libya GeoTec (vol 4)
Zeleriakoval M, Purcz P, Hlavata IG (2012) Comparison of precipitation trends in Libya and Slovakia. River Basin Manag VII 172:365
Zhou S, Yu B, Zhang Y (2023) Global concurrent climate extremes exacerbated by anthropogenic climate change. Sci Adv 9(10):p.eabo1638
Zurqani HA, Mikhailova EA, Post CJ, Schlautman MA, Elhawej AR (2019) A review of Libyan soil databases for use within an ecosystem services framework. Land 8(5):82
Zurqani HA (ed) (2021) The soils of Libya. World soils book series, 1st edn. Springer, Cham. ISBN-13: 978-3030663674
Zurqani HA, Al-Bukhari A, Aldaikh AO, Elfadli KI, Bataw AA (2022) Geospatial mapping and analysis of the 2019 flood disaster extent and impact in the city of Ghat in Southwestern Libya using Google Earth engine and deep learning technique. In: Zurqani HA (eds) Environmental applications of remote sensing and GIS in Libya. Springer, Cham. https://doi.org/10.1007/978-3-030-97810-5_10

Dr. Hamdi A. Zurqani is an Assistant Professor of Geospatial Science in Natural Resource Management and Conservation at the University of Arkansas Agricultural Experiment Station, Arkansas Forest Resources Center, University of Arkansas at Monticello, USA. With a strong academic foundation and a passion for innovative technologies, he is also an FAA (i.e., Pt107) Certified sUAS/Drones Pilot, leveraging this expertise to enhance his proficiency in remote sensing, GIS, and geospatial data analysis, which he applies to advance his research and teaching in environmental monitoring and analysis. Dr. Zurqani is a recognized expert as a result of his internationally acclaimed work in the areas of environmental information science, remote sensing, geospatial analysis, land evaluation, sustainability, pedology, and soil science education. He has conducted research across the world, including the United States of America and Africa, and has served as PI, co-PI, or co-investigator on several grants-funded research projects. In addition to his research and teaching responsibilities, Dr. Zurqani is actively involved in institutional service at the University of Arkansas at Monticello. He serves on the UAM Campus-wide Artificial Intelligence (AI) Review Committee, the UAM Faculty Assembly Student Engagement Committee, and the UAM Library Committee.

Professionally, Dr. Zurqani contributes to the broader academic community through a variety of leadership and collaborative roles. He is an active member of the Steering Committee of the Institute of Electrical and Electronics Engineers (IEEE) Mediterranean and Middle-East Geoscience and Remote Sensing Symposium (M2GARSS) and serves as an Ambassador for the Geoscience and Remote Sensing Society (GRSS) Chapter of the IEEE in Libya, facilitating communication, collaboration, and knowledge-sharing within the geospatial and remote sensing communities. In recognition of his expertise and substantial contributions to the field, Dr. Zurqani was invited in 2024 to serve on the Technical Program Committee for the IEEE Asia-Pacific Conference on Geoscience, Electronics, and Remote Sensing Technology (AGERS). In 2025, he also joined the IEEE GRSS Technical Committee on Remote Sensing Environment, Analysis, and Climate Technologies (REACT), a platform for scientists and engineers to exchange ideas and knowledge on environmental issues and the impacts of climate change, emphasizing support for sustainable development goals (SDGs) through remote sensing technologies. Beyond these roles, Dr. Zurqani holds the position of UAM Co-Principal Investigator at the Southeast Climate Adaptation Science Center, where he is involved in research initiatives focused on climate adaptation and resilience in the southeastern United States and its surrounding regions.

Dr. Zurqani's editorial and peer-review activities further highlight his commitment to advancing scientific research. He serves on the Editorial Boards of several journals, including *Frontiers in Earth Science* (Geoinformatics), *Discover Soil* (Springer Nature), *Frontiers in Remote Sensing*, and *Remote Sensing* (MDPI). He has also been appointed as a Guest Editor for special issues in several journals, focusing on topics such as advancing spatial prediction of soil properties using remotely sensed data and geospatial artificial intelligence, and UAV agricultural management. Dr. Zurqani's dedication to his field is further demonstrated by his role as a peer reviewer for numerous publishers, including Springer Nature, Nature, Frontiers, Elsevier, and MDPI, among others. He is also the editor of several influential publications that address critical environmental challenges. His notable works include: *The Soils of Libya* (2021), part of Springer's World Soils Book Series; *Environmental Applications of Remote Sensing and GIS in Libya* (2022); and *Water Resources of Libya: Challenges and Management* (2025), published as part of the Springer Water Series (SPWA). These publications highlight his multidisciplinary expertise and commitment to addressing top environmental issues in Libya and beyond.

Dr. Abdulsalam Al-Bukhari, Ph.D. is an Assistant Professor of the Natural Resources Management, Remote Sensing, and Geographic Information Systems at the Department of Forestry and Rangeland, Faculty of Natural Resources and Environmental Sciences, Omar Al-Mukhtar University, Al Bayada, Libya. He received his M.Sc. in Environmental Sciences (rangeland ecology and management) in 2010 from Omar Al-Mukhtar University, Al Bayada, Libya. Dr. Albukhari completed his Ph.D. in Data Risk and Environmental Analytical Methods (The Use of Remote Sensing and GIS for Monitoring Rangeland Degradation in Libya) from Cranfield University, Cranfield, UK, 2019. He specialized in rangeland ecology and management. His main research interests focused on monitoring and assessing rangeland degradation and the environmental information sciences (Remote Sensing and GIS). He is head of the Geographic Information Systems Unit at Omar Al-Mukhtar University. Dr. Albukhari is a member of the Editorial Board for Al-Mukhtar Journal of Sciences.

Ms. Shadia A. Alzurqani is a Science Geology graduate from the University of Tripoli, Libya, and she is currently pursuing her graduate studies at the University of Arkansas at Monticello, USA. Within the College of Forestry, Agriculture, and Natural Resources. Her ongoing graduate project focuses on monitoring the spatial and temporal effects of drought on vegetation, employing remote sensing technology, mainly Google Earth Engine (GEE). During her academic pursuit at

UAM, she published a peer-reviewed paper presented at the prestigious 2023 I-GRASS Conference held in Pasadena, California. Furthermore, her research findings were presented at many other conferences, such as the 2023 South Dakota Student Water Conference.

Dr. Mhi Eldin M. Alkhboli, Ph.D. is a Full Professor at the Department of Soil and Water Sciences at the Faculty of Agriculture, University of Omar Al-Mukhtar, Al Bayada, Libya. He completed his M. Sc. in Soil Physics in 1996 from the University of Omar Al-Mukhtar, Libya, and earned his Ph.D. in Environmental Studies in 2003 from the University of Alexandria. Dr. Alkhboli's primary research and teaching focus includes land use management, land evaluation, remote sensing, and geographical information systems. He has worked at various departments, including the Department of Soil and Water, Faculty of Agriculture, Faculty of Natural Resources, Department of Geography, Faculty of Art, University of Omar Al-Mukhtar (2003–2013), Department of Science, Environmental Engineering and Department of Urban and Regional Planning, Graduate Academy Benghazi Branch (2010–2023). Dr. Alkhboli provided technical assistance in establishing the GIS Unit at the Agricultural Research Center in Tubruk (2007) and Al Bayada (2010). He also worked as part of the Water harvesting team for the joint project between ICARDA and Agricultural Research Center, Libya (2010–2012). Dr. Alkhboli has several scientific research papers published in peer-reviewed scientific journals and conferences and has supervised numerous master's theses in the field of interest. Additionally, Dr. Alkhboli was commissioned by the Ministry of Higher Education and Scientific Research to establish a Date Palm Tree Research Center and serves as its general director (2013–2024).

Dr. Ali A. Bataw, Ph.D. is a professor of Animal Ecology at the Department of Zoology, Faculty of Sciences, University of Omar Al-Mukhtar, Libya. He got his M.Sc. (Agricultural Sciences) in 1985 from University of Tripoli, Tripoli, Libya. Dr. Bataw completed his Ph.D. in (Behaviour Ecology) from St. Andrews University, UK, in 1995. He specialized in Insect Ecology. His main research interests focused on wild bees taxonomy, flowers-bees interaction, biodiversity, and insect ecology. He has been teaching courses and has supervised the graduation of master's and bachelor's degree students since 1996.

Future of Water Supply and Demand in Libya Under Climate and Land-Use Change

Ismail Ageena, Nagib Froja, and Salah Hamad

Abstract Climate change is a significant global concern with widespread consequences for ecosystems, agriculture, and water resources in various regions. This chapter focuses on the relationship between climate patterns and water supply and demand in Libya over 60 years, from 1961 to 2020. The research reveals a significant warming trend, particularly between July and October, with notable temperature increases in several locations. Precipitation in Libya is erratic, and evapotranspiration exhibits regional variations. Water availability is a critical concern due to its arid climate and limited freshwater resources, and agriculture is the main consumer of water. Sustainable land management practices are crucial in addressing climate change, with deforestation and urbanization contributing to around 23% of global greenhouse gas emissions. This chapter also provides valuable insights into climate change trends in Libya and emphasizes the importance of understanding these trends for informed decision-making and strategies to mitigate the environmental and socio-economic impacts of climate change.

Keywords Climate change · Trend analysis · Temperature · Precipitation · Evapotranspiration · Water consumption · Land use

---

I. Ageena
Department of Geography, Faculty of Arts, University of Al-Zawiya, Al-Zawiya, Libya
e-mail: i.ageena@zu.edu.ly

N. Froja (✉)
Soil and Water Department, Faculty of Agriculture, University of Tripoli, Tripoli, Libya
e-mail: N.froja@uot.edu.ly

S. Hamad (✉)
Faculty of Natural Resources and Environmental Science, Omar Al Mukhtar University, Al Bayda, Libya
e-mail: salah.hamad@omu.edu.ly

H. A. Zurqani (ed.), *Water Resources of Libya*, Springer Water,
https://doi.org/10.1007/978-3-031-80920-0_9

1 Introduction

Climate change is undeniably one of the most crucial global challenges of the twenty-first century, drawing considerable attention and concern (IPCC 2018; Steffen et al. 2018; Stocker et al. 2013). To monitor this phenomenon, more than 11,000 weather observation stations worldwide continually collect data on diverse weather and climatic parameters, including those over land and sea (IPCC 2007; Trenberth et al. 2015). This extensive network of measurement is complemented by data from satellites, ships, and aircraft, providing a comprehensive understanding of climate dynamics (Trenberth et al. 2016). The World Meteorological Organization (WMO) has played a significant role in climate data collection, with records dating back to the late seventeenth century, including monthly temperature, precipitation, and air pressure data (Brönnimann et al. 2019). These records are essential for tracking long-term climate patterns and discerning the impacts of climate change (Folland et al. 2001; Braganza et al. 2013). Natural scientists have progressively shifted their focus to climate change during the latter part of the twentieth century, recognizing its far-reaching implications for the environment (IPCC 2018; Solomon 2007; Knutti et al. 2017). This increased scientific attention underscores the urgency of understanding and addressing this complex issue. The first World Climate Conference in 1979 marked a pivotal moment, leading to the establishment of the WMO World Climate Programme. This initiative aimed to coordinate efforts to address climate change and enhance international collaboration on climate science. In 1988, the Intergovernmental Panel on Climate Change (IPCC) was founded, reflecting a significant international response to the challenge of climate change (Bolin 2007). The IPCC has since released a series of assessment reports, starting in 1990, which provide comprehensive evaluations of climate science and its implications. Subsequent reports have been published in 1995, 2001, 2007, 2013, and 2021, significantly advancing our knowledge of climate change dynamics. The awareness of environmental dangers posed by pollution began to rise in the 1950s, coinciding with the initial emergence of evidence linking human activities to rising global temperatures (IPCC 2018). This increased awareness has been a driving force behind efforts to address climate change and its causes.

Temperature is one of the most easily measured climatic variables, and changes within the climate can be directly assessed, with series generally longer than for other climatic variables. The implications of climate change for temperature are substantial in the Mediterranean and North Africa regions, given the potential sensitivity of agriculture and population to temperature changes and the pressure placed on water resources (Loaiciga et al. 2000; Gonzales 2001; Huntingford et al. 2005; Kafle and Bruins 2009). Studies examining global, regional, and localized temperature changes have increased over the last couple of decades due to the increasing recognition of the spatial variability in the rates of change (Hansen et al. 2001; Della-Marta et al. 2007), regional (e.g., Ben-Gai et al. 1999; Hasanean 2001; Xoplaki et al. 2003; Bartolini et al. 2008; Bermejo and Ancell 2009), and local studies (El-Tantawi 2005; ElKenawy et al. 2009).

Precipitation is vital for agricultural systems and the provision of potable water within Libya. Droughts, with serious risks to human lives and properties, are rapidly changing (temporally and spatially) across North Africa. The IPCC (1996, 2001a, 2013, 2021g) warned that global changes in precipitation regimes and warming would lead to increases in both floods and droughts. Global and regional precipitation trends have increased over the last couple of decades due to the increasing interest in the potential impacts and responses to climate change.

Evapotranspiration is the second-largest flux in the water cycle, and any changes would affect the entire water cycle. Evapotranspiration is an important climatic parameter controlling surface energy exchange, energy transport, and transformation in the global atmospheric system. Increases in global warming led to higher surface temperatures, which may generate more evaporation. The fluctuations in temperature and precipitation trends resulting from climate change are expected to have a strong effect on the spatial and temporal distribution of water resources (IPCC 2007), and thus impact patterns of evaporation. Changes in climate and extreme weather are considered the most serious environmental challenge that threatens developed and developing countries. Libya, one of the driest countries in the world, has recorded significant changes in temperature and precipitation during the last couple of decades, with serious impacts on the environment, human welfare, and socio-economic systems. This issue is being taken seriously by international authorities and has received considerable attention from the Libyan government.

This chapter aims to comprehensively analyze the climate change trends in Libya, with a particular focus on temperature, precipitation, and evapotranspiration patterns from 1961 to 2020. The investigation seeks to shed light on the environmental and socio-economic implications of these changes. With Libya being one of the world's driest countries, these trends have significant consequences for water availability and agricultural systems, which are crucial for the well-being of the population. Moreover, the study emphasizes the importance of understanding these climate change dynamics for informed decision-making and the development of strategies to mitigate the associated environmental and socio-economic challenges. By addressing water availability issues and promoting sustainable land management practices, we can take essential steps toward tackling this pressing global issue and safeguarding the future of water supply and demand in Libya.

2 Climate of Libya

2.1 Temperatures

The climate of Libya is characterized by a mix of subtropical and Mediterranean climates which is interposed between the subtropical desert (hot, dry summers) and temperate maritime (mild, relatively wet winters) climate. In addition, the climate of coastal Libya is characterized by wet winters and hot dry summers and shows a transition between the Mediterranean and temperate climates to the arid climates of the interior. The climate of Libya is influenced by a variety of topographical and geographical parameters, with varying significance within the two (coastal and inland) regions, with considerable temporal and spatial variability in temperature identified across Libya (Fig. 1). The mean annual average temperature across Libya is 20.8 °C, with an annual average range between 14.2 to 21.0 °C at Northern (costal) stations and between 21.3 and 23.4 °C at Southern (inland) stations during the period 1961–2020 (Ageena 2013). The highest temperature throughout the world (57.8 °C) was observed on 13 September 1922 in Al-Aziziya city, with the lowest recorded temperature of − 8.3 °C recorded at Ghadames on 11 January 1935. This study focuses on data recorded for the period 1961–2020, as dataset integrity improved considerably after 1961 at the majority of coastal stations.

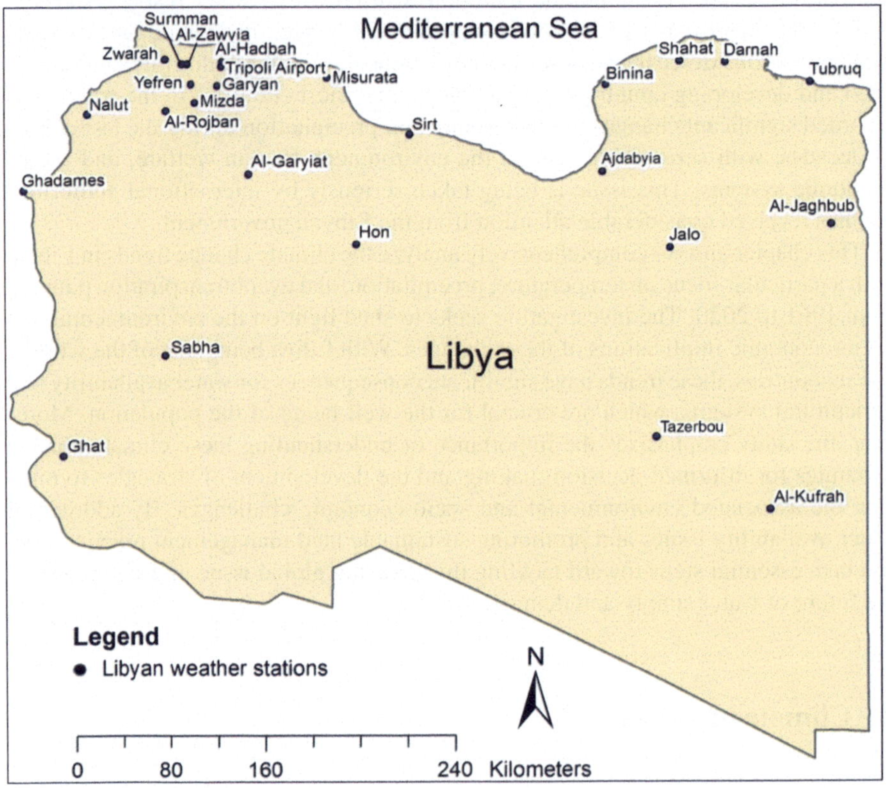

Fig. 1 The distribution of stations within this study across Libya

2.1.1 Mean Monthly Average Temperature

Positive changes (increases) in mean monthly average temperature are found at most stations; with positive trends ranging from 0.001 °C/year to 0.10 °C /year, with significant positive trends identified July–October at most stations, particularly Sirt, Ajdabyia, Ghadames, Al-Garyiat, Al-Jaghbub, Sabha and Tazerbou (Table 1). In September significant increases are identified at all stations except Nalut, Musratah, and Binina, with highly significant increases (***) at Sirt and Ajdabyia, with significant (**) increases at Shahat, Hon, Sabha and Tazerbou (Table 1). In February, negative trends in mean monthly average temperature ranges between 0.012 and − 0.001 °C /year (Musratah, Sirt, Binina, Shahat, Darnah, Al-Garyiat, Hon, Jalo, Tazerbou and Al-Kufrah). In order to examine differences in monthly average temperature across Libya, the variability of the mean monthly average temperature of 15 stations for two 30 years' intervals 1961–1990 and 1991–2020 were analysed to show the variability of monthly average temperature during the last 30 years compared to the 30 first years (1961–1990).

Changes in mean monthly average temperature for the two periods (1961–1990 and 1991–2020) show higher average temperature at all stations, except Musratah, Binina and Jalo in the second period 1991–2020. The Mann–Whitney test results show, that the warm months (July–October) had much higher temperatures, significant at 95% for the period 1961–2020 at all stations, with the except at Musratah, Binina and Jalo.

2.1.2 Annual Variations and Trends in Mean Average Temperature

The mean annual average temperature for the period 1961–2020 exhibits a slight increase across the 15 stations (with Al-Garyiat, Ghat and Tazerbou excluded), with a clearer increase by the mid-1980. The Mann–Kendall results identify positive trends in mean annual average temperature at all stations (Table 2), with an average of 0.021 °C/year, with the highest 0.043 °C/year (Hon). Significant trends in mean annual average temperatures are identified at twelve of 15 stations, with three exceptions Musratah, Jalo and Al-Jaghbub. Highly significant increases (***) are identified at stations: Zwarah, Ajdabyia, Ghadames, Hon, Sabha and Al-Kufrah, with significant increases (**) at Tripoli Airport, Sirt, Darnah and (*) at Nalut, Binina, Shahat. With No significant change at Musrath, Al-Garyiat, Jalo, Al-Jaghbub, Ghat and Tazerbou (Table 2).

Analysis of the data from 1961 to 2020 was undertaken by dividing the whole period in two, providing two periods for analysis of 30 years in length, this permits consideration of general long-term changes and differences in rates of change. Decreasing (increasing) mean average temperature is identified at eight (seven) of 15 stations, with an average rate of -0.015 °C/year (0.012 °C/year) for the first 30 years (1961–1990) (Table 2), with statistically significant decreases at Musratah (***) and Jalo (**). Increases in annual mean average temperature are identified at all (18) stations over the last 30 years (1991–2020), with an average rate of 0.049 °C/

Table 1 Statistically significant differences in temperatures between the periods 1961–1990 and 1991–2020 at weather monitoring stations in Libya

Stations	Jan	Feb	Mar	Apr	May	Jun	Jul	Aug	Sep	Oct	Nov	Dec
Zwarah	0.039***	0.026**	0.032***	0.031***	0.032***	0.017**	0.032***	0.042***	0.042***	0.052***	0.036***	0.035***
T. Airport	0.014	0.006	0.022*	0.021*	0.024*	0.006	0.029**	0.029**	0.025**	0.038**	0.019*	0.015+
Nalut	0.038**	0.013	0.028*	0.024*	0.017	0.014	0.024*	0.026**	0.020*	0.050***	0.026*	0.027*
Musratah	0.007	0.002	0.013	0.006	0.004	– 0.013	– 0.001	– 0.009	0.013+	0.022*	0.011	0.013*
Sirt	0.012*	0.001	0.015+	0.017**	0.015+	0.017*	0.020*	0.023***	0.37***	0.027**	0.014	0.012*
Ajdabyia	0.020**	0.008	0.018+	0.016*	0.021*	0.006	0.030***	0.040***	0.045***	0.023*	0.016+	0.016*
Binina	0.012+	0.001	0.017	0.009	0.009	0.006	0.001	0.11+	0.026***	0.018+	0.005	0.014*
Shahat	0.009+	– 0.009	0.011	0.013	0.001	0.014*	0.20**	0.014*	0.023**	0.020+	0.003	0.000
Darnah	0.009+	– 0.001	0.008	0.011	0.021**	0.008	0.014*	0.013*	0.020**	0.022**	0.001	0.012
Ghadames	0.017	0.009	0.029*	0.022*	0.035**	0.003	0.032**	0.050***	0.041**	0.041**	0.034*	0.021+
Hon	0.022	– 0.012	0.017	0.033	0.028*	0.008	0.085***	0.070***	0.089***	0.100***	0.076**	0.039+
Jalo	0.022	– 0.012	0.024	0.020*	0.024	0.008	0.033**	0.046***	0.078**	0.019**	0.074	0.037+
Jaghbub	0.013+	0.000	0.001	0.001	0.007	0.006	0.030***	0.015+	0.047***	0.013	0.004	0.006
Sabha	0.020	0.006	0.029*	0.035**	0.040**	0.017+	0.034**	0.053***	0.051**	0.041*	0.021	0.027+
Al-Kufrah	0.019*	0.010	0.021+	0.043***	0.029*	0.022**	0.038***	0.026**	0.046***	0.021+	0.028*	0.021*

Table 2 Annual mean average temperature, with statistically significant levels across Libya

Time series	1961–2020			1961–1990			1991–2020		
	Temp	Sig	Q	Temp	Sig	Q	Temp	Sig	Q
Zwarah	20.4	***	0.028	20.2		– 0.012	20.6	***	0.056
T. Airport	20.7	**	0.021	20.5		0.000	20.9	***	0.040
Nalut	19.4	*	0.024	19.2		0.012	19.7	***	0.052
Musrath	20.5		0.005	20.2	***	– 0.043	20.7	***	0.076
Sirt	20.7	**	0.016	20.4		0.012	21.0	***	0.064
Ajdabyia	20.9	***	0.031	20.7		0.003	21.1	***	0.058
Binina	20.1	*	0.029	19.9		0.004	20.4	***	0.680
Shahat	16.8	*	0.033	16.5		0.012	17.1	*	0.560
Darnah	20.6	**	0.021	20.4		0.032	20.9	***	0.032
Ghadames	22.0	***	0.028	21.7		– 0.033	22.3	**	0.067
Al-Garyiat	#			#			21.5	**	0.049
Hon	21.6	***	0.034	21.2		– 0.014	21.9	***	0.045
Jalo	22.8		0.006	22.4	**	– 0.021	22.9	***	0.049
Al-Jaghbub	21.3		0.008	21.1		– 0.015	21.6	***	0.051
Sabha	23.1	***	0.034	22.9		0.013	23.4	***	0.047
Ghat	#			#			24.5	**	0.034
Tazerbou	#			#			23.1	***	0.072
Al-Kufrah	23.6	***	0.024	22.9		0.000	23.9	***	0.069

year, with the highest 0.076 °C/year (Musratah). These increases are significant at all 18 stations, with highly significant (***) at 14 stations, Shahat (*), Al-Garyiat, Ghadames, and Ghat (**). Increases (13 stations) or no change (Musratah and Jalo, with Al-Garyiat, Tazerbou and Ghat excluded) in average temperature are identified when comparing between the two timeframes (1961–1990 and 1991–2020), with increases identified in period two compared to period one. Significantly higher (95% confidence level) mean annual average temperatures are identified at nine stations; Zwarah, Tripoli Airport, Nalut, Sirt, Ajdabyia, Ghadames, Al-Jaghbub, Sabha and Al-Kufrah.

2.2 *Precipitation*

Precipitation is a highly unpredictable weather variable across North Africa and is rapidly changing (temporally and spatially) affecting populations and societies across North Africa. The IPCC (1996, 2001a) warned that global changes in precipitation regime and warming will lead to increases in floods global. The lack of precipitation is the prevailing characteristic across much of Libya, with approximately 93% of the total land surface of Libya receiving less than 100 mm/year, with precipitation rare, irregular, and reducing progressively towards zero from north to south.

2.2.1 Temporal Variability of Precipitation

The Libyan rainy season is defined as follows: autumn (September, October and November), winter (December, January and February) and spring (March, April and May), with ~ 99% of the total annual precipitation occurring within these months. Precipitation is generally characterized by high spatial and temporal variability. The mean annual precipitation along the Libyan coast ranges between 140.0 mm (Nalut; 150 km from the sea) and 557.0 mm (Shahat; 10 km from the sea); with few years exceeding 50.0 mm in southern Libya. This section studies a temporal and spatial variability in precipitation based on monthly and annual data during the period 1961–2020 (Table 3).

The coefficient of variation (COV) of mean total annual precipitation at coastal stations is 63.0%, ranging from 22.7% (Shahat; which has the highest total precipitation in the eastern coastal region, ~ 10 km from the coast, its elevation 621 m to 58.3% Gharyan, which has the highest total precipitation in the west coastal region, 81 km from the coast its elevation 891 m). The COV of the mean total annual precipitation at inland stations ranges between 63.5% (Al-Garyiat) and 184% (Al-Kufrah). However, variability of precipitation indicates more stability at coastal stations than inland stations, indicating that the main factor influencing spatial variability in COV mostly to its proximity to the Mediterranean Sea (Table 3 and Fig. 2).

The annual precipitation for the 14 western stations (Table 3) over the last 63 to 81 years (depending on length of the individual station record), illustrates that most of the annual precipitation ranges between 100 and 300 mm (Fig. 2), representing approximately 69% of the years within the series, with considerable spatial variability, with 90% at Nalut and 45% et al.-Hadbah of years receiving below 300 mm (Table 3). The eastern stations, receive annual precipitation between 100 and 300 mm during 52% of the years studied, with less than 300 mm received during only 3% of years at Shahat and 72% of years at Tubruq (Table 4). The annual total precipitation for inland stations indicates that 88% of the years receive < 50 mm, with 26% at Mizda, 10% et al. Garyiat, 2.6% at Ghadames and Hon.

2.2.2 Monthly Variation and Trends

The rainy season can be characterised as increasingly heavy rain until January and decreases thereafter (except at Nalut which has a peak in precipitation during March), where December and January are the wettest months with the six months of October to March receiving approximately 87% of the total annual precipitation (Ahmed 2002). By the mid of January, almost 75% of annual total precipitation has fallen at coastal stations (eastern western, and central regions). Meanwhile, the western mountains region receives 75% of the total rainfall in February. The mean monthly

Table 3 Percentage of annual normal precipitation for each interval meteorological stations across Libya for 1961–2020

Precipitation	< 100 mm	100–200	201–300	301–400	401–500	> 500
Zwarah	7.9	35.5	38.2	13.2	5.3	0.0
Surmman	0.0	30.5	44.4	13.8	11.1	0.0
Al-Zawyia	5.5	25.9	35.1	24.0	7.4	1.9
T. Airport	1.3	17.5	47.5	22.5	8.8	2.5
Al-Hadbah	0.0	6.0	39.4	33.3	18.2	3.0
Yefren	3.6	21.4	39.3	28.6	7.1	0.0
Garyan	0.0	16.6	27.3	27.2	12.2	16.6
Al-Rojban	15.2	34.7	32.6	13.0	4.3	2.1
Nalut	30.0	56.3	11.3	1.3	0.0	1.3
Musrath	0.0	23.9	46.3	28.4	0.0	1.5
Sirt	12.5	45.0	26.3	7.5	2.5	0.0
Ajdabyia	21.0	59.0	16.1	3.2	0.0	0.0
Binina	0.0	24.2	40.9	27.3	7.6	0.0
Al-Bayda	0.0	3.3	0.0	15.6	18.7	62.5
Shahat	0.0	3.2	0.0	2.9	19.4	64.5
Darnah	0.0	24.6	47.7	21.5	6.2	3.2
Al-Fatayah	3.2	12.9	25.8	38.7	16.1	0.0
Tubruq	7.6	65.3	26.9	0.0	0.0	0.0
Mizda	63.3	36.6	0.0	0.0	0.0	0.0
Ghadames	98.7	0.0	1.3	0.0	0.0	0.0
Al-Garyiat	88.1	11.9	0.0	0.0	0.0	0.0
Hon	97.5	2.5	0.0	0.0	0.0	0.0
Jalo	100.0	0.0	0.0	0.0	0.0	0.0
AlJaghbub	100.0	0.0	0.0	0.0	0.0	0.0
Sabha	100.0	0.0	0.0	0.0	0.0	0.0
Ghat	100.0	0.0	0.0	0.0	0.0	0.0
Tazerbou	100.0	0.0	0.0	0.0	0.0	0.0
Al-Kufrah	100.0	0.0	0.0	0.0	0.0	0.0

precipitation is roughly symmetrical around the mid-season peak and has a sinusoidal shape for most coastal stations, with non-symmetrical distributions identified in the western mountain region and all inland stations (Fig. 3).

The estimation of changes in monthly precipitation for 13 stations (Table 3; with the remaining stations excluded as they have shorter records or/and low monthly amounts of rainfall) during the period 1961–2020 have been analysed using the nonparametric Mann–Kendall test, which is used to estimate the statistically significant of precipitation trends within the series. Mean monthly precipitation data shows

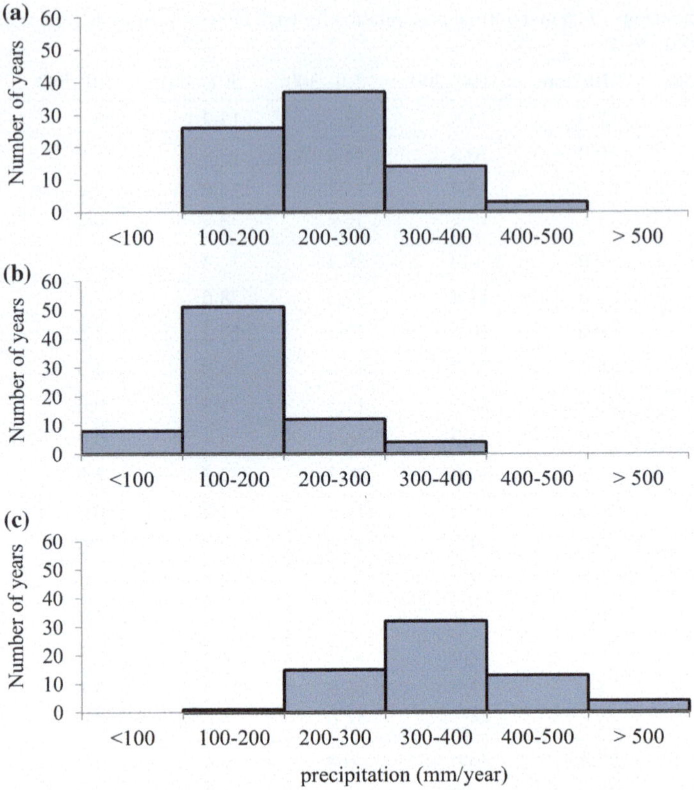

Fig. 2 Histograms of mean annual total precipitation at coastal **a** western; **b** central; **c** eastern stations across Libya

that, the cold months, November-March are more variable, while the driest moths are more stable with no changes in monthly precipitation identified; April (62%), May and September (69%) of stations during the period 1961–2020. Positive increases in monthly precipitation are identified in December (54%) with an average rate of 0.141 mm/month, with no statistically significant trends at any stations. About 54%, of stations show increasing trend in January, ranging between 0.006 and 0.537 mm/month, with significant increases at Musratah (*) and Zwarah (+). Increases at 84% (62%) of study stations in February (March), with an average rate of 0.137 (0.066) mm/month, with significant increase (*) at the central stations of Sirt and Ajdabyia (Table 3).

According to previous studies of climate, that refers to a change in climate at the beginning of the seventies of the last century; increases of temperature in Libya during the last 30 years. However, analyses of precipitation variation during the two periods are applied in this study. The study period (1961–2020) is divided into two series of equal length (30-year), 1961–1990 and 1991–2020, referred to as period 1 and period 2, respectively to examine differences in monthly precipitation for 13

Table 4 Mean monthly precipitation (MMP), with statistically significant levels at stations across Libya for 1961–2020

Stations		Sep	Oct	Nov	Dec	Jan	Feb	Mar	Apr	May
Zwarah	Q	0.029	– 0.188+	0.121	0.109	0.246	0.14	– 0.005	0.000	0.000
	M MP	13.5	35.0	38.1	46.3	36.4	22.1	17.3	12.3	5.6
Al-Zawyia	Q	0.000	– 0.22	0.409	0.455	0.537+	0.252	0.038	0.000	0.000
	M MP	15.0	41.0	41.4	54.9	50.0	29.0	23.2	10.4	4.1
T. Airport	Q	0.000	– 0.205+	– 0.094	– 0.371	0.026	0.069	0.007	0.000	0.000
	M MP	10.8	35.4	40.3	53.7	54.4	33.0	25.8	15.4	4.4
Gharyan	Q	0.000	– 0.044	– 0.078	0.057	0.356	0.041	0.000	– 0.125	0.000
	M MP	9.7	43.0	29.1	53.5	61.9	42.6	55.5	20.6	11.7
Nalut	Q	0.003	0.000	– 0.031	0.039	– 0.091	0.011	0.138	– 0.045	0.000
	M MP	5.3	18.6	14.0	18.5	16.1	19.2	26.0	16.5	10.7
Musratah	Q	0.01	– 0.231	0.239	– 0.15	– 0.04	0.064	0.082	0.011	0.000
	M MP	11.3	37.9	45.9	58.4	56.5	29.0	21.8	9.8	4.5
Sirt	Q	0.009	0.007	0.03	0.017	0. 394*	0.257*	0.053	0.000	0.000
	M MP	10.0	23.3	24.5	43.4	38.8	23.2	15.2	4.5	3.0
Ajdabyia	Q	0.000	0.000	0.025	0.156	– 0.08	0.267*	0.048	– 0.015	0.267
	M MP	1.5	9.1	18.9	44.7	39.1	20.4	11.1	3.4	2.5
Binina	Q	0.000	– 0.059	0.003	– 0.281	– 0.035+	0.22	0.095	0.000	– 0.004
	M MP	3.2	19.4	34.5	66.3	66.0	40.9	25.9	6.3	4.4
Shahat	Q	0.000	– 0.068	– 0.206	– 0.226	– 0.626	0.26	– 0.052	0.000	– 0.006
	M MP	9.2	52.2	68.5	116.3	123.8	87.5	66.9	22.7	8.9
Darnah	Q	0.000	– 0.121	– 0.07	0.151	0.238	0.173	– 0.066	0.000	0.005
	M MP	5.6	34.7	28.7	56.8	60.1	39.6	23.6	8.3	5.7

(continued)

Table 4 (continued)

Stations		Sep	Oct	Nov	Dec	Jan	Feb	Mar	Apr	May
Ghadames	Q	0.000	0.010	0.000	0.000	0.000	0.000	0.046	0.000	0.000
	M MP	1.6	3.6	1.7	5.2	5.9	5.3	5.9	2.9	1.8
Hon	Q	0.000	− 0.013	0.000	0.000	0.006	0.000	0.000	0.000	0.000
	M MP	2.6	5.2	3.6	2.6	4.5	3.0	3.3	2.9	3.2

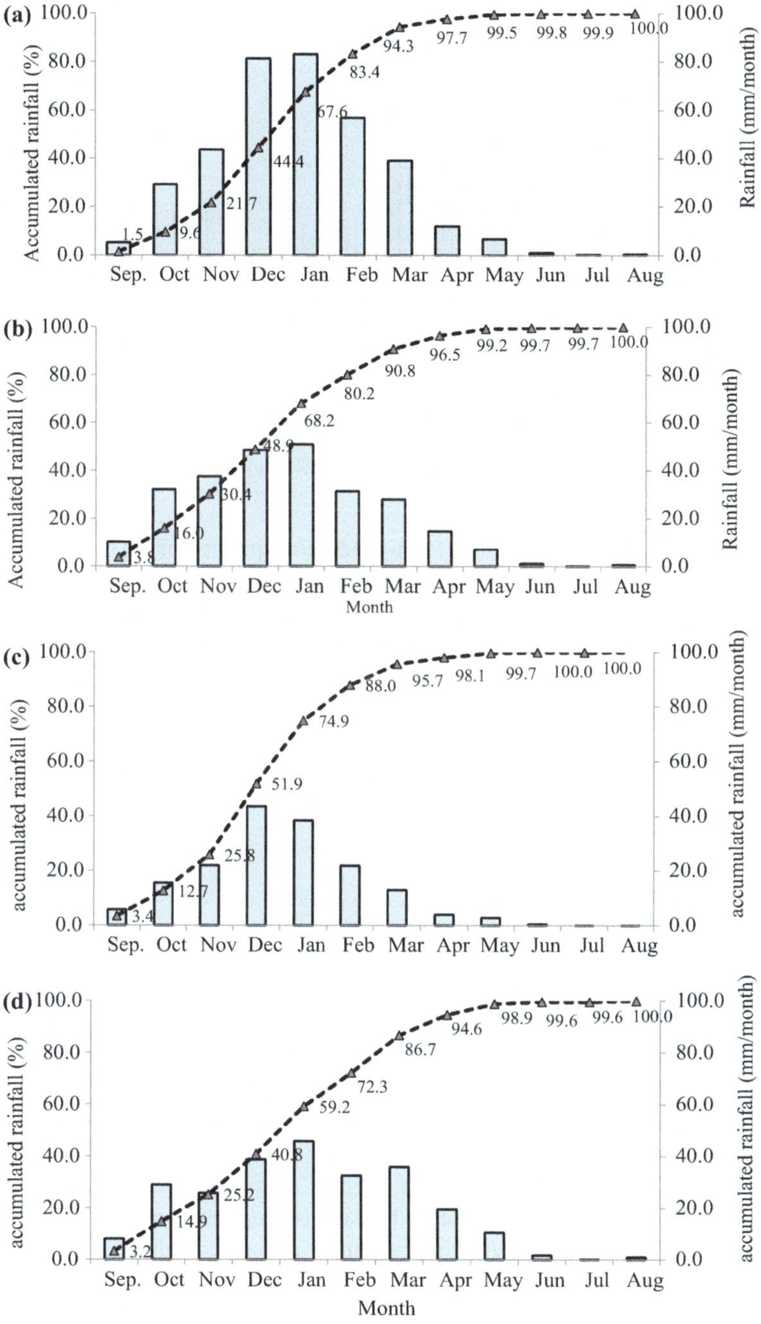

Fig. 3 Histogram and cumulative monthly precipitation (1961–2020), at; **a** eastern-coastal region; **b** western-coastal region; **c** central-coastal region and; **d** western mountains region

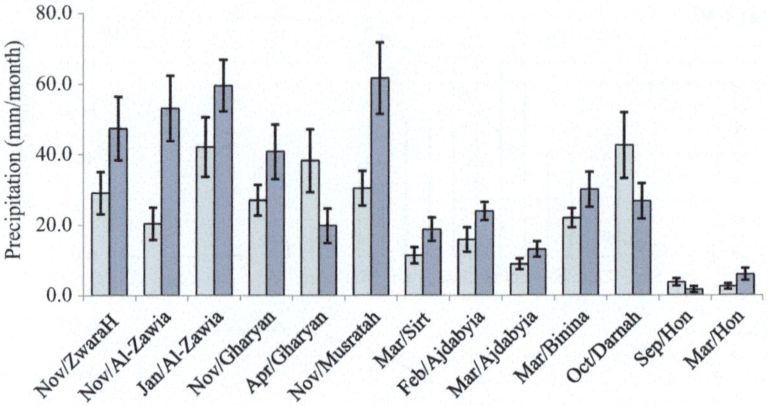

Fig. 4 Mean monthly precipitation for the 30-year intervals of two periods 1961–1990 and 1991–2020 for significant months with one standard error (68.2% confidence level)

stations across Libya over the period 1961–2020. Comparison of the two time periods for monthly precipitation does not show clear or/and little significant difference based on Mann–Whitney test (U-test; Wilcoxon, 1945) (confidence level 95%) for most of months except September (Nalut, Musratah, Sirt), December (Ghadames), April (Hon) and May (Shahat and Darnah). Error-bars (Fig. 4) show that differences (95% significance level) in monthly precipitation between the two periods 1961–1990 and 1991–2020 in September (Hon), October (Darnah), November (Zwarah, Al-Zawia, and Musratah), January (Al-Zawia), Febrarury (Ajdabyia), March (Sirt and Hon), April (Garyan) and May (Sirt, Ajdabyia and Binina).

2.2.3 Annual Variations and Trends

Since precipitation is highly variable both temporally and spatially across Libya, the 28 stations are used to properly capture precipitation patterns, during the different time periods: 1961–2020; 1961–1990 and 1991–2020. Changes and trends in annual total precipitation for 16 stations during the period 1961–2020 (60 years) are analysed (with the remaining stations excluded as they have shorter records; Table 5). The Mann Kendal test identifies positive increases in annual precipitation (1961–2020) at seven of the ten coastal stations ranging between 0.052 and 0.402 mm/year (Table 5. 5). Negative trends are found at Tripoli Airport, Binina and Shahat, with weakly significant at Tripoli Airport (+). The four inland stations examined identify positive trends (0.104 mm/year), with a significant increase et al.-Jaghbub (*). A negative trend is found at Hon, with no decrease trends at inland stations (Table 5).

The annual total precipitation for 16 stations across Libya for the first 30-year period (1961–1991) is analysed (with the remaining stations excluded as they have shorter records). Analysis of the variation in precipitation patterns reveals positive trends in annual precipitation at six coastal stations (1.998 mm/year), with the highest

Table 5 Statistically significant levels for annual precipitation at metrological stations across Libya

Station	Annual total	COV %	1961–2020		1961–1999		1991–2020	
			sig	Q (mm/y)	sig	Q (mm/y)	sig	Q (mm/y)
Zwarah	222.6	38.2		0.167		2.742		− 1.961
Surmman	237.4	34.7						− 0.444
Al-Zawia	256.2	43.4						− 1.879
T. Airport	267.2	36.4	*	− 1.025		2.443	*	− 2.972
Al-Hadbah	309.5	29.3						− 3.433
Yefren	333.5	59.1		0.052		4.982	*	− 4.74
Garyian	258.6	30.1						− 0.123
Al-Rojban	217.4	56.1						− 7.35
Nalut	138.2	53.2		0.104		− 0.348	**	− 1.5
Musratah	248.9	29.7		0.152		0.224		− 1.576
Sirt	167.4	44.3		0.402		1.818		− 0.295
Ajdabyia	146.3	42.8		0.628	*	1.711		0.333
Binina	256.3	31.6		− 662		− 1.505	*	− 2.561
Al-Bayda	523.4	42.2						3.833
Shahat	549.1	26.8		− 1.761		0.89		0.08
Darnah	254.3	31.5		0.278		− 0.632		0.054
Al-Fatayah	306.2	37.2						− 0.017
Tubruq	165.7	28.4						− 0.911
Ghadames	32	92.6		0.1		− 0.311		0.3
Mizda	59.6	68.2						1.217
Al-Garyiat	54.2	93.4						0.371
Hon	28.6	67.4		− 0.015		− 0.147		− 0.396
Jalo	9.4	90.6		0.001		0.094		− 0.113
Al-Jaghbub	13.2	115.7	*	0.107	*	0.359		− 0.362
Sabha	7.4	110.5		0.001		0.001		0.125
Ghat	8.4	114.7						0.355
Tazerbou	2.2	76.8						− 0.022
Al-Kufrah	1.5	186.4		0.001		0.001		0.001

(4.982 mm/year) and with significant increases (+; 1.711 mm/year at Ajdabyia). Negative trend is found at four coastal stations (− 0.843 mm/year). At inland stations, Ghadames and Hon were decreases trends (− 0.229 mm/year), with the remaining inland stations were positive, with a significant increase (+) at Jalo (Table 5).

The annual precipitation at all 28 stations for the last 30 years (1991–2020) across Libya are analysed; the Mann–Kendall test identifies decreasing trends at 14 of the coastal stations (78% of coastal stations; − 2.126 mm/year), with positive trends at Ajdabyia and the eastern stations of Al-Bayda, Shahat and Darnah (1.075 mm/year).

Significant decreases are identified et al.-Rojban (*), Tripoli Airport and Tazerbou (*) and yefren (+). At inland stations, positive (negative) trends 0.395 mm/year (−0.225 mm/year) are identified at six (four) stations, with a significant increase at Ghat (**) and decreases et al.-Jaghbub (+) and Tazerbou (**) (Table 5).A time series of mean annual total precipitation over the two 30 year intervals for 16 stations across Libya (1961–1990 and 1991–2020) are analysed to provide a clear depiction of annual precipitation change (with the remaining stations excluded as they have shorter records). Changes in mean annual total precipitation show that period 2 is characterized by higher precipitation compared to the period 1, at 43% of the stations, with no changes at Zwarah and Binina. Significant differences in mean annual total precipitation are found when comparing between the two timeframes, with the second period significantly higher (68.2% confidence level) at Tripoli Airport, AlZawia and Al-Jaghbub.

2.2.4 Spatial Variability of Precipitation

Libya is divided into five geographical regions; Al-Jfarah and Nafusa Mountains, Gulf of Sirt, plain of Benghazi and Al-Jabal Al-Akhdar, the North Sahara, and South Sahara (Fig. 5). In general, precipitation decreases gradually towards the centre of the Libyan coastline (near Sirt) and decreases sharply with latitude, reaching nearly zero in the south-east of Libya. The mean average annual total precipitation at the eastern region is 350.8 mm/year, at western stations 253.7 mm/year and at the central stations 168.1 mm/year, with the lowest annual total precipitation in the Sahara region 22.3 mm/year. The highest annual total precipitation was 555.4 mm/year (Shahat) followed by 534.1 mm/year (AlBayda) and 345.6 mm/year (Gharyan), with the lowest precipitation in the coastal region recorded at Nalut (147.6 mm/year) followed by Ajdabyia (149.2 mm/year). The mean total annual precipitation in the northern Sahara region does not exceed 62.6 mm/year at Mizda, with the mean annual total precipitation across the southern Sahara region 10 mm/year.

2.3 Evapotranspiration

Evapotranspiration is the second largest flux in the water cycle thus any changes, would affect the whole water cycle. Evapotranspiration is the most difficult and complicated component of the water cycle (Xu and Singh 2005) and is also a very important indicator of climate change (Peterson et al. 1995; Brutsaert and Parlange 1998). In the surface water balance, approximately 60–80% of the precipitation on the terrestrial surface returns back into the atmosphere, where it becomes the source of future precipitation (Tateishi and Ahn 1996). Evapotranspiration is an important climatic parameter controlling surface energy exchange, energy transport, and transformation in the global atmospheric system, with increases in global warming leading to get higher surface temperatures, which may generate more evaporation

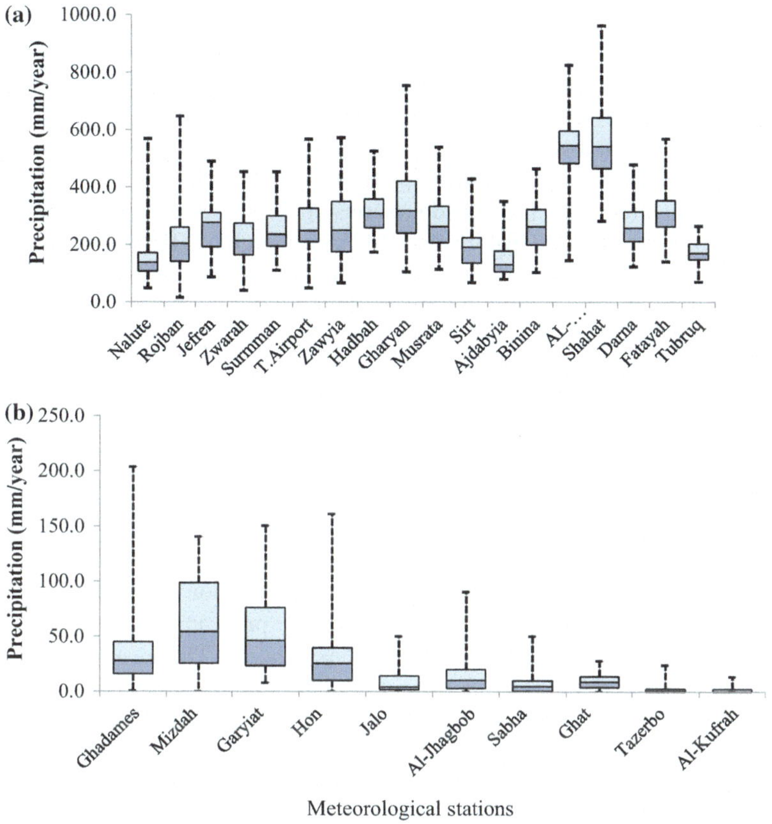

Fig. 5 Annual total precipitation for the 28 stations across Libya (1961–2020); **a** coastal stations; and **b** inland stations, with maximum, minimum, lower quartile (first quartile—25%), median (50%) and the upper quartile (third quartile—75%)

(Ageena and Froja 2020). The fluctuations of temperature and precipitation trends resulting from changes in climate are expected to have a strong effect on the spatial and temporal distribution of water resources (IPCC 2007) and thus impact patterns of evaporation. Evaporation occurs from water surfaces, whether lakes, rivers, or reservoirs, and it is called (lake evaporation). Evaporation also occurs from the surface of the earth and is called (soil evaporation), and from trees and plants, it is called transpiration. Therefore, the term evapotranspiration is used to express the sum of evaporation and transpiration together. Potential evapotranspiration (PET) is defined as the maximum amount of water removed from the surface through the processes of evaporation and transpiration when water availability is not limited.

In this section, spatial and temporal characteristics of Potential evapotranspiration (PET) are analyzed across Libya based on the Penman–Monteith (PETP-M) method which is recommended by the Food and Agricultural Organisation of the United Nations (FAO). The Penman–Monteith method (Jensen et al. 1990) requires data for the following variables for calculation: sunshine duration (hours), humidity (percentage) and wind speed (km), minimum and maximum temperature (°C). The geographic density of stations allowing for ET calculation is restricted to the 16 synoptic stations distributed across Libya during the period of study (1961–2020).

The Mann–Kendall test (MK); (Mann 1945; Kendall 1975), was applied to detect the significance of trends in evapotranspiration data; the MK test, is more suitable where monotonic trends are found in data series of 1945–2010; (Zhou and Ren 2011; Ageena 2013; Feng et al. 2016). Inverse Distance Weighted (IDW) tool in ArcMap was used as an Interpolation tool to predict the Evapotranspiration trend values of cells at locations that lack sampled points. It involves using known z values and weights determined as a function of distances between the unknown and known points. As such in IDW points that far away have far less influence than points that are close (Childs 2004).

2.3.1 Observation, Changes, and Trends of Potential Evapotranspiration (PETP-M)

The mean annual and seasonal of potential evapotranspiration based on the Penman–Monteith equation (using the CropWat 8.0 software program based on Allen et al. 1998; Raja 2010) was estimated for 16 stations (Fig. 6). The mean annual potential evapotranspiration (PET) at the meteorological stations across Libya during the period 1961–2010 ranges between 1509.4 (Darnah) and 2054.1 (Nalut) mm/year at the Northern (costal) stations, and from 2128.2 (Al-Jaghbub) to 3026.3 mm/year (Sabha) for the Southern (inland) stations, with the mean average of PETP-M estimated by 1703.5 mm/year and 2501.3 mm/year at the Northern and Southern stations, respectively (Table 6).

Table 1 shows the highest seasonal of PETP-M in summer, with an average rate ranging between 33 and 39% of annual total evapotranspiration with the lowest in winter (13%). The mean seasonal PETP-M of spring (28%) and autumn (23%) are almost similar.

Linear regression of PETP-M (1961–2020) was performed on monthly and annual estimates. For Northern and Southern regions, a spatially averaged time series was analyzed from all stations (Fig. 6). Increases trends (Table 6) are identified at four Northern stations as average (2.733 mm/year), with significant (95% confidence level) trends at Zwarah, Musratah, Bininaand Ajdabyia. The mean linear trends for the southern region illustrate decreases (– 7.101 mm/year) in annual PETP-M at four stations, with the highest value (– 11.173 mm/year) at Tazerbou, with significant (95% confidence level) decreases et al.-Jaghbub, Ghadmes, Jalo, and Tazerbou. Increasing trends are identified at three Southern stations (2.879 mm/year), with no positive significant trend at any Southern stations. Several studies have shown that the development of irrigation in arid zone oasis agriculture has resulted in a reduction in wind speed and an increase in relative humidity. As a consequence of these changes, reference crop evapotranspiration in this region has decreased (Han et al. 2010).

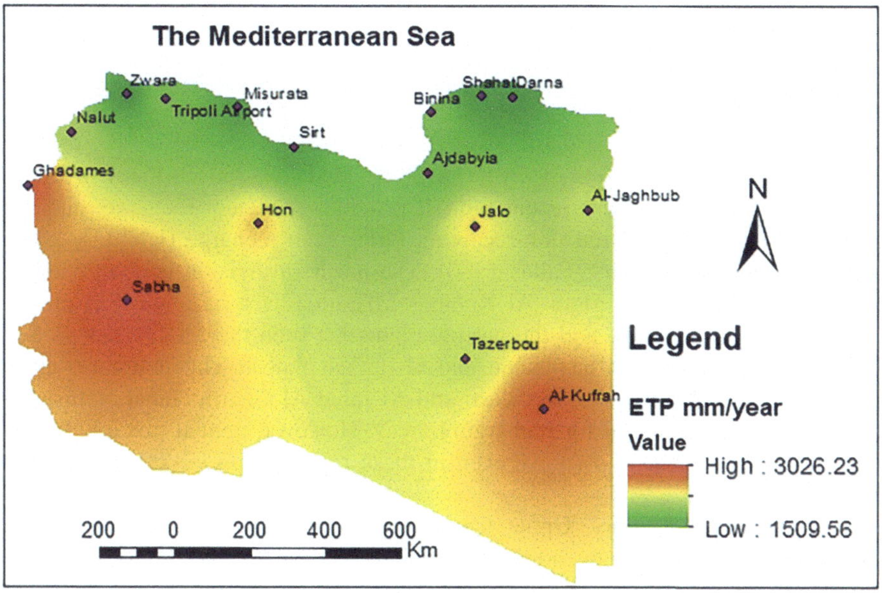

Fig. 6 The mean annual evapotranspiration (mm) (1961–2010)

Table 6 Absolute values of annual and seasonal PETP-M (mm/year) for the 16 stations and Trend rate, with significant (95% confidence level) for the period (1961–2020)

Time series	Trend rate	Annual	Autumn	Winter	Spring	Summer
Zwarah	2.867*	1512.4	373.8	245.4	380.5	512.7
T. Airport	− 0.448	1851.3	416.7	219.3	493.0	722.2
Nalut	− 5.555*	2054.1	462.3	264.1	547.2	781.2
Musratah	2.899*	1585.6	381.1	240.6	412.2	551.7
Sirt	− 0.131	1599.5	389.9	251.7	424.8	533.2
Ajdabyia	3.655*	1813.1	412.2	217.3	520.3	663.3
Binina	1.511	1865.3	443.4	217.2	517.3	687.3
Shahat	− 2.960*	1541.0	343.8	181.1	407.6	608.5
Darnah	− 0.223	1509.4	365.2	245.9	399.3	519.0
Ghadames	− 6.834*	2757.0	619.8	304.1	763.2	1069.9
Hon	3.400	2375.3	529.4	301.6	684.4	860.0
Jalo	− 1.636	2332.4	513.0	306.0	665.6	847.8
AL-Jaghbub	− 8.759*	2128.2	453.4	257.7	612.4	804.7
Sabha	3.640	3026.3	697.2	363.6	868.5	1096.9
Tazerbou	− 11.173 *	2133.1	479.4	285.8	636.0	731.9
Al-Kufrah	1.598	2756.5	641.1	363.3	779.6	972.5

3 Water Availability

3.1 Water Resources

Conventional water resources, Libya can be divided into the northern division, consisting of aquifers where recent rainfall provides sufficient recharge, and the southern division, considered non-receiving little or no recharge (UNECA 2005; Bakhbakhi and Salem 1999; Pallas 1980). The northern division is comprised of the coastal plains; Jifarah plain, Al Khomes Misrata-Tawergha plain, Sirt plains, Ajdabiya-Benghazi plains, and the narrow plains between the Mediterranean Sea and Al Jabal Al Akhdar, Dafna plateau, and Al-Batnan plateau. The mountain ranges of Al Jabal Al Akhdar in the northeast, and Al Jabal Al Gharbi (Jabal Nafusa) in the northwest are essential for recharging areas. However, annual precipitation is typically less than 500 mm (mm), even in these regions (FAO 2009; El-Tantawi 2005).

The Quaternary, Tertiary, Upper Cretaceous, and Triassic aquifers are where groundwater is found in the northern division. Current rainfall recharges these aquifers (Pallas 1980). Even so, there are some areas of central Libya's coastline that can have saline groundwater, as well as the coastal region of Tobrock, which stretches from the Gulf of Bambah to the Egyptian border. Due to sufficient recharge, the Benghazi plain, the Jifarah plain, and Al Jabal Al Akhdar have high-quality groundwater, particularly in sub-coastal areas (Wheida 2012; Hamad 2012; Salem 1992; Shaki 2006). The majority of the sandstone aquifers in the southern parts of Libya are from the Tertiary, Lower Cretaceous, Triassic, and Palaeozoic eras (Hamad and Ahweej 2020; Binsariti and Saeed 2000; Salem 1992; Pallas 1980). These aquifers are a significant source of high-quality water with a good salinity. The waters are of late Pleistocene and Holocene age (Guendouz and Moulla 2010; Swailem et al. 1984), Also, these aquifers are not currently being recharged, and are therefore non-renewable.

Libya has five groundwater management divisions (Groundwater systems) (see Fig. 3.6, Chap. 3) and (Table 7) identified by Pallas (1980), Jones (1971); Jifarah plain, Gadamis-Al Hamadah, Murzuq, AI Jabal Al Akhdar, Sirt-Sarir and AI Kufrah. These groundwater divisions include areas that are not strictly geological basins but have been erroneously indicated as groundwater basins in some sources.

Furthermore, as illustrated in (see Fig. 3.7, Chap. 3), with its neighbours in North Africa (Tunisia, Algeria, Egypt, Sudan, Chad, and Niger), Libya shares four trans-boundary aquifer systems, including the Djeffara Aquifer System (extension of the Jifarah system in Libya), Lake Chad Basin (extension of the Murzuq system), North-west Sahara Aquifer System (NWSAS) (extension of the Gadamis-Al Hamadah system in Libya), and Nubian Sand (Hamad and Saaid 2022; Ahmed 2020; Frédéric 2020).

Unconventional water resources are mainly the desalination used since the early 1960s in Libya (NPC 2014). The General Company for Desalination (GDC) is presently the primary supplier of desalinated water for domestic uses. The General

Table 7 Available Conventional water resources Million cubic metre (NPC 2014)

Groundwater systems	Groundwater renewable	Groundwater non-renewable	Surface water	Total
Jifarah	300	50	25.5	375.5
Gadamis-Al Hamadah	50	350	16.4	416.4
Murzuq	0	2500	0	2500
AI Jabal Al Akhdar	400	100	375	875
Sirt-Sarir	0	1570	3.4	1573.4
AI Kufrah	0	1650	0	1650
Total	750	6220	420.3	

Company for Water and Wastewater (GCWW) provides desalinated water through eight desalination plants connected to distribution networks. The GDC was established in 2007 to operate and maintain eight desalination plants along Libya's coastline; these plants are planned to supply about 134 urban settlements along or near the Libyan coasts, with no alternative water supply sources. The total installed capacity of these desalination plants is 350,000 m^3/day, a quantity that is sufficient to satisfy the demands of about 1.4 million inhabitants or about one-fourth of Libya's population (UNICEF 2019; NPC 2014). Due to many challenges the GDC has encountered, the amount of water supplied has significantly decreased over the last few years; the total amount of water produced in 2017 averaged only about 98,000 m^3/d, indicating a production efficiency of only 28%. To put it another way, only 28% of the time did the served population have access to desalinated water (UNICEF 2019). For the rest of the "interrupted supply time", they had to fetch it from other "more expensive private" sources. Although, desalination is becoming a solution for water scarcity in most arid countries, yet Libyan government not adopted desalination solution as a strategic option to solve water scarcity. Wastewater treatment plants were mostly designed for producing water suitable for agricultural use. The largest operating wastewater treatment plants are located in Tripoli, Misurata and Sirte with a design capacity of 110,000, 24,000 and 21,000 m^3/day respectively. Furthermore, most of the remaining wastewater facilities are medium and small sized plants (370–6700 m^3/day). The estimated amount of wastewater to be treated is 1324million m^3/day, and the estimated amount of treated wastewater is about 145,800 m^3/day which accounts for only 11% and the remaining percentage of wastewater is being pumped into the sea, artificial lagoons, and black wells without any treatment (Brika 2018).

3.2 *Water Demand*

Water balance in Libya limited to precipitations and groundwater abstractions. And since the early of seventies of the last century, Libya has faced a serious water shortage problem. United Nations classified Libya as one of the ten poorest countries in its

Table 8 Projected Libya's water balance (LMP 2019)

Year	1995	2000	2005	2010	2015	2020	2025
Supply (Mm^3)	3820	3820	3820	3820	3820	3820	3820
Demand (Mm^3)	3885	4493	5128	5794	6495	7236	8022
Balance (Mm^3)	– 65	– 673	– 1308	– 1974	– 2675	– 3416	– 4202

renewable water resources and has experienced the situation of imbalance between water resources and water needs (GWA 1999). Regarding to National Water Security Strategy report published on 2014. Agriculture had the highest consumption quantity is about 2680 million m^3/year which represents (83%) of the total water consumption, while the domestic and industrial sectors withdrew about (12%) and 5% respectively. The total available fresh water supplies on sustainable basis have been estimated at the fixed rate of 3820 million m^3per year. In contrast, water demand has grown from 3882 million m^3 in 1995 to 5794 million m^3 in 2010 and expected to be 8022 million m^3 by 2025 (Table 8).

The gap between water supply and demand can be reduced by improving water use efficiency both in agriculture and in domestic use. The challenge in this issue is the shift from managing water supply to managing water demand. This switch will become essential to balance Libya's water budget and sustainable face future water shortages, reducing the danger of potential water crisis that may arise (LMP 2019; Al-Samarrai 2018).

3.3 Implications of Change in Climate on Environmental Variables and Water Availability

Climate change will have the greatest effect on the agricultural sector, as a result of warming and significant temporal and spatial changes in precipitation, which will increase rates of water loss through evapotranspiration, this led to the drying up of some dam lakes, for example the complete drying up of the water reserved in front of the Wadi Kaam Dam in Al-Khoms region (Fig. 7a, b), note that the average annual volume of water in the Wadi Kaam Dam Lake is 13 million cubic meters (El-Baruni 2022).

The sixth report of the Intergovernmental Panel on Climate Change (2022) indicates that the Mediterranean basin region is one of the regions most vulnerable to climate change, and a hot spot for climate extremes such as heat waves and drought waves, which have become more frequent and severe. The report also indicates that the southern bank of this basin will witness a decrease in Rainfall will reach 40% during the current decade, and these areas will suffer severe water stress, affecting water supplies and food security. Moreover, an increase in some other climate extremes such as floods, flash floods, droughts and heat waves in severity and magnitude. Climate change is already causing the growing season to start earlier,

(a) October.2020 (b) Augst.2021

Fig. 7 Drought of Wadi Kaam dam (*Photo courtesy* El-Baruni)

lengthening the growing season leading to earlier flowering, especially in warm and moderate regions, such as the region around Gharyan at Jabal Nafusa in western Libya and in Al-Jabal Al-Akhdar in eastern Libya. Changes in the frequency, length and timing of heat waves in the agricultural region will affect agricultural yields and increase the number and variety of insects. The heat waves that swept the Northwest of Libya in the past two years (2020, 2021) which lasted for more than ten continuous days, led to the deterioration of the production of many crops.

Precipitation is an important factor in Libya for agricultural purposes, particularly in the north region, irrigated using groundwater from northern aquifers, which are recharged by precipitation. Changes in frequency, length, amount and number of precipitation days have significant effects on agricultural yields and change the growing season. Increased temperatures and decreased precipitation lead to an increased rate of evapotranspiration, which will result in a greater demand for groundwater for agricultural processes. Decreases in precipitation could put freshwater resources at risk, for example, interference of seawater with coastal aquifers as a result of the significant decrease in rainfall in recent years. The expected climate changes will have strong impacts on water resources, especially since these resources suffer from many restrictions and pressures that make them in a fragile situation and are more affected by the warming and drought that may occur in the future (Elfadli 2019). Water scarcity is endemic in the country and changes in the water balance would have substantial implications for agriculture and population, increased desertification and associated socio-economic impacts. In recent years (2020–2023), Jabal Nafusa in the Northwest of Libya has witnessed a severe decrease in the rate of rainfall for several consecutive years, which has resulted in a significant decline in the rainfed agricultural area and the drought and death of many of the fig and olive trees (Figs. 8 and 9).

The appearance of large unusual water eddies belonging to the family of marine cyclones, which are a circular body of water on some Libyan coasts, especially during the past ten years, which indicates a dramatic change in sea surface temperature rates near the Libyan coasts. For example, these water eddies appeared 4 times on

Fig. 8 Drying olive trees in Nalut (field photos taken by: Froja, 24.2. 2022)

Fig. 9 Drying fig trees in Nalut (field photos taken by: Froja, 24.2. 2022)

the beaches of Tripoli and Tajoura on January 2015, and they also appeared 3 times during January 2020 on the beaches of Zuwara and in September 2022 in Al-Bayda and Derna (Elfadli 2019). These eddies can develop and turn into hurricanes and devastating floods under the influence of climate change. Flooding is one of the most important climate change risks to human life in both rainy regions and regions that rarely have significant precipitation, particularly in the large metropolitan area where floods destroy homes and leave thousands of people homeless. Frequently, thousands of people living around Wadi Al-Mjanin are forced to leave their homes because of flooding as a result of unexpected precipitation (e.g. 1986). Moreover, flooding can cause overflows from sewage treatment plants and impact human health reducing the availability of fresh water and food. As extreme precipitation events are expected to become more intense in the future, these events are likely to lead to more risk. For example, on 10.9.2023 torrential rains from Storm Daniel led two dams close to the city of Derna to burst, pushing entire neighborhoods into the sea (see Fig. 3, Chap. 8). This devastating storm comes in an unprecedented year of climate disasters and record-breaking weather events, from devastating wildfires to excessive heat waves. The floods that swept Ghat region during the period (June 1–6, 2019) (see Fig. 2, Chap. 8), which are considered among the most severe and unprecedented floods in at least 130 years. It was also repeated on October 1, 2023, on Ghat and Ubari cities in the far southwest of Libya, near the Algerian border.

4 The Relationship of Climate Change with Land Uses

4.1 The Impact of Land Use on Climate Change

The intricate interplay between land use changes and climate change is a pivotal facet of environmental science, bearing profound implications for the future of our planet. Land use alterations exert a substantial role in driving climate change by emitting greenhouse gases into the atmosphere (Shukla et al. 2019). This includes the grievous consequences of deforestation and the transformation of natural ecosystems into agricultural or urban areas, which result in the release of carbon dioxide (CO_2) and methane (CH_4)—two prominent greenhouse gases. These activities are estimated to account for approximately 23% of global greenhouse gas emissions (Shukla et al. 2019). However, the implementation of sustainable land management practices, such as reforestation and afforestation, possesses the potential to ameliorate the impacts of climate change by sequestering CO_2 and bolstering ecosystem resilience (Shukla et al. 2019). These practices hold the capability to effectively counterbalance the greenhouse gas emissions associated with land use changes. Moreover, land use changes yield considerable regional consequences on climate, influencing the intensity, frequency, and duration of extreme weather events (Shukla et al. 2019). These impacts are contingent upon location and season, underscoring the intricate nature of the relationship between land use and climate change. Furthermore, these alterations in land conditions may trigger far-reaching repercussions, such as the escalation of dust storms in arid areas, with detrimental ramifications for human health (Shukla et al. 2019).The repercussions of land use changes also extend to the realm of water resources, where urbanization and afforestation play a pivotal role in shaping water retention and nutrient export. For instance, augmenting forest cover within a catchment area can reduce water resources due to amplified evaporation from the headwaters (Gallart and Llorens 2003). It is imperative to factor in these fluctuations in land cover and land use when striving to manage water resources in a sustainable manner. In certain cases, the amalgamated effects of land use changes and climate change on various ecosystem services can be more profound than their individual impacts (Bai et al. 2019). This accentuates the urgency of fostering a holistic understanding of these interactions to enlighten policy formulation and land management decisions. To confront the environmental and climatic challenges of our era, it is essential to instate effective forest management and urban planning. Sustainable land use practices are paramount in mitigating the consequences of climate change and in adapting to rising sea levels (Shukla et al. 2019). In a nutshell, the intricate and bidirectional relationship between land use and climate change underscores the imperativeness of efficacious land management and conservation practices for the sustainability of ecosystems and natural resources.

Land Use and Climate change are deeply interconnected (Dale 1997), influencing each other at various temporal and spatial scales. However, improper land uses primarily drive climate change (Turner 1994; Turner and Meyer 1994). This paper synthesizes relevant literature and the case of Nepal to assess the relationship between land use and climate change (Thapa and Dhulikhel 2019) while focusing on how land-use impacts climate change and vice versa. In recent centuries, land-use change has significantly affected ecological variables and climate change (Freeman 2015). Understanding research on both subjects will guide decision-makers and conservation planners in sustainable land and climate management (Schmidt et al. 2020). In the past few decades, land use and climate change have emerged as two prominent global concerns, necessitating comprehensive research to understand their intricate relationships and future implications. Nonetheless, the causes and consequences of human-induced climate change and land-use activities have seldom been studied in isolation (Dale 1997; Turner 1994; Turner and Meyer 1994). Researchers and stakeholders recognize their close interdependence (Freeman 2015; Schmidt et al. 2020) and advocate for a concerted multidisciplinary effort to explore this process, encompassing both socioeconomic and natural science (Dirmeyer et al. 2010; Hibbard et al. 2010). Land-use change is the process by which human activities transform the natural landscape (Peng et al. 2017). It refers to the alterations in land use driven by economic activities (Rudel 2009). Land-use change is catalyzed by factors such as population growth, deforestation, changes in food consumption, and the escalating demand for energy and fiber (Dirmeyer et al. 2010; Hibbard et al. 2010). Climate change encompasses global warming, shifts in precipitation patterns, and the emergence of natural disasters like floods, storms, and droughts (Jamet and Corfee-Morlot 2009; Seddon et al. 2020). Land-use activities influence climate by causing deforestation and the rapid expansion of human populations, while climate change shapes land use through unpredictable heavy rainfall and rising temperatures (Dirmeyer et al. 2010; Hibbard et al. 2010). Climate change's impacts on land use are manifest in changes in driving forces that fluctuate across time and space. Land-use change (LUC) fundamentally underpins global adjustments and directly influences climate change (Riebsame et al. 1994). This review encompasses current practices in assessing land-use change in relation to climate change (Turner and Meyer 1994). Developing countries bear the brunt of changes in climate and land use, grappling with limited capacity to prevent and respond to their impacts (Dirmeyer et al. 2010; Hibbard et al. 2010; Bellard et al. 2012). These limitations magnify the devastating consequences and exacerbate vulnerabilities within local communities. By approximately 2030, climate change is projected to inflict a substantial toll on the poorer segments of the population, leading to increased mortality, malnutrition, the spread of tropical diseases, and urban heat stress (Patz et al. 2005). Additionally, climate and land-use changes contribute to the occurrence of natural disasters, a phenomenon substantiated by several studies, particularly in developing countries with constrained resources, technology, and budgets for disaster mitigation and resilience (Dawson and Spannagle 2008).

4.2 Libya Land Use Change and Climate Change Impacts

Land use change in Libya, especially in its urban areas such as Tripoli and Benghazi, has witnessed significant transformations over the years. Rapid urbanization and expansion, driven by population growth and economic development, have led to profound alterations in land cover. These changes are often driven by the conversion of natural landscapes into urban areas characterized by increased impervious surfaces like roads and buildings (Abuelgasim et al. 2016). One notable consequence of this land use change is the exacerbation of the urban heat island (UHI) effect. The UHI effect is characterized by higher temperatures within urban areas compared to their rural surroundings. As a result of increased impervious surfaces and reduced vegetative cover, urban areas in Libya experience elevated local temperatures (El-Hattab et al. 2017). This not only impacts the comfort and health of urban residents but also contributes to higher energy consumption for cooling, further intensifying the greenhouse gas emissions that drive climate change.

Additionally, the conversion of natural landscapes to urban areas in Libya has disrupted the ecological balance. This leads to decreased vegetation cover, which plays a crucial role in regulating temperature and precipitation patterns. These changes can disrupt local microclimates and affect precipitation patterns (Hussin and Abdullah 2018). In some cases, land use changes may reduce soil infiltration capacity, contributing to localized flooding during heavy rains, which poses a direct threat to urban infrastructure and residents (Farhan and Elgarni 2018). Moreover, the transformation of land from agricultural use to urban development may also have repercussions for food security. As arable land diminishes due to urban expansion, it can impact food production capacity and the availability of locally sourced food (Tellería et al. 2019). Furthermore, the desertification process, which refers to the spread of arid and semi-arid conditions, is exacerbated by land use changes. As natural landscapes are converted into urban areas, it can lead to the further expansion of arid regions, aggravating environmental challenges in Libya (Libyan Environmental General Authority 2017). Understanding the intricate relationship between land use change and climate change is pivotal for sustainable urban development in Libya. This knowledge can guide policy and planning efforts to mitigate the adverse impacts of land use changes and promote more resilient urban development (Abuzied and Pradhan 2017; Fekri and De Maeyer 2019). Measures to counteract these effects may include urban planning strategies that prioritize green spaces and sustainable development practices to mitigate the UHI effect, conserve natural ecosystems, and secure food production and water resources in a changing climate.

5 Conclusion

This chapter discusses the relationship between climate patterns and water supply and demand in Libya over 60 years, from 1961 to 2020, with a particular focus on temperature, precipitation, and evapotranspiration. The results show that climate change has led to significant shifts in temperature, precipitation, and evapotranspiration, with profound implications for Libya's environment, water resources, and socio-economic systems. Temperature analysis reveals substantial increases in coastal regions, particularly from July to October, and an overall upward trend in annual temperatures. Precipitation patterns are irregular, with changing trends in monthly and yearly precipitation. Potential evapotranspiration (PET) varies significantly by region, with higher PET values in northern coastal areas. The chapter emphasizes the importance of sustainable water management and efficient utilization of water resources to bridge the gap between supply and demand. Human-induced land use changes significantly affect greenhouse gas emissions, regional climates, and food security. Sustainable urban development practices, such as integrating green spaces and efforts to combat the urban heat island effect, are necessary to promote resilient urban development in the face of a changing climate.

References

Abuelgasim A, Alsadig I, Bokhari N (2016) Evaluation of land use and land cover change using remote sensing and GIS: case study of Tripoli, Libya. J Geographic Inf Syst 8(01):1–10

Abuzied MS, Pradhan B (2017) Urban growth analysis and land use/land cover change detection in Tripoli city, Libya. Arabian J Geosci 10(20):434

Ageena I (2013) Trends and patterns in the climate of Libya (1945–2010); Thesis submitted in accordance with the requirements of the University of Liverpool for the degree of Doctor of Philosophy. University of Liverpool

Ageena I, Froja M (2020) Temporal and spatial Analysis of Evapotranspiration in Libya. Libyan J Agric 24(2)

Ahmed M (2002) Climate variability and its impacts. J Environ Stud 12:34–49

Ahmed M (2020) Sustainable management scenarios for Northern Africa's fossil aquifer systems. J Hydrol

Allen RG, Pereira LS, Raes D, Smith M (1998) Crop evapotranspiration: guidelines for computing crop water requirements. FAO Irrigation and Drainage Paper. No. 56.

Al-Samarrai M (2018) Climate change challenges in the Arab world. Clim Policy Stud 5(2):67–80

Al-Samarrai KA, Sadeg SA (2018) Precision irrigation efficient technologies practices in Libya from the water and energy point of view. Int J Appl Nat Sci (IJANS) 9(6)

Bai Y et al (2019) Global warming impacts on biodiversity. J Ecol 107(2):423–432

Bakhbakhi M, Salem O (1999) Why the great man-made river project? In: General Water Authority & the L. A. Jamahiriya (eds) Regional aquifer systems in arid zones—managing non-renewable resources. UNESCO, pp 3–16

Bartolini G, Meloni D, Mastrorillo L (2008) Observed climate variability and trends in central Mediterranean climate (Italy) over the last century. Nat Hazard 8(5):1247–1258

Bellard C et al (2012) Impacts of climate change on future biodiversity. Ecol Lett 15(4):365–377

Ben-Gai T, Bitan A, Manes A, Alpert P (1999) Long-term changes in annual and seasonal rainfall in the eastern Mediterranean, 1950–1998. Theoret Appl Climatol 64(1–2):109–121

Bermejo V, Ancell R (2009) Climate change in Iberia: state of the art. Clim Change 97(1–2):19–35
Binsariti A, Fawzi Saeed (2000) Groundwater salinity variations in the Cambro Ordovician Aquifer of Eastern Jabal Al Hasawnah, the Great Man-Made River Project, Libya. In: Sola DWMA (ed) Geological exploration in Murzuq Basin. Elsevier Science B.V., pp 1–16
Bolin B (2007) A history of the science and politics of climate change: the role of the Intergovernmental Panel on Climate Change. Cambridge University Press
Braganza K et al (2013) Nature of climate variability. Int J Clim Stud 33(4):601–620
Brika B (2018) Water resources and desalination in Libya: a review. Proceedings 2(11):578. https://doi.org/10.3390/proceedings2110578
Brönnimann S, Annis JL, Dann W, Krämer D, Stickler A (2019) Historical weather data for climate research: rescue and digitization. Clim Res 77(2):95–107
Brutsaert W, Parlange MB (1998) Hydrological processes under climate change. Water Resour Res 34(8):2297–2303
Childs C (2004) Interpolating surfaces in ArcGIS spatial analyst. ArcUser 32–35
Dale VH (1997) The relationship between land-use changes and climate change. Ecol Appl 7(3):753–769
Dawson RJ, Spannagle M (2008) The potential for retrofitting sustainable drainage systems: a review. Water Environ J 22(1):13–22
Dawsons RJ et al (2010) Impacts of land use change on climate. Wiley Online Library
Della-Marta PM, Haylock MR, Luterbacher J, Wanner H (2007) Doubled length of western European summer heat waves since 1880. J Geophys Res Atmospheres 112(D15)
Dirmeyer PA et al (2010) Land surface climate feedback mechanisms. Clim Dyn 34(1):117–129
El-Baruni S (2022) Climate change in Libya and its impact on the environment and water resources. The seventh conference of the College of Arts (climate changes in Libya), Sirt University
Elfadli K (2019) Drought risk and mitigation in the Libyan desert. J Arid Environ 23(5):419–432
El-Fadli KI (2019) Potential effects of climate change on drought in Libya. PhD. thesis, Cairo University, Egypt
ElKenawy AM et al (2009) Trends in climate data in Egypt. Int J Climatol 29(3):455–472
El-Tantawi AM (2005) Trends of temperature and precipitation in Upper Egypt. Egypt J Remote Sens Space Sci 8(1):39–47
ElKenawy AM, And JK (2009) Climate change and its impact on the Nile Delta, Egypt. Environ Geol 58(1):163–168
El-Hattab MM, Salama MS, Pimbert MP (2017) Analysis of Urban Land use and cover change and its impact on climate in Tripoli Metropolitan City, Libya. Remote Sens 9(2):176
Elsheikh R, Al-Fugaira A, El-Gaaly G (2014) Remote sensing and GIS-based analysis of land use/land cover change in the coastal region of Libya. J Coast Conserv 18(1):83–95
FAO (2009) FAO report on climate chang e and food security. Food and Agriculture Organization
Farahat E, Al-Bakoush O (2018) Evaluation of land use and land cover changes in Tripoli, Libya. Open J Geol 8(10):1150–1166
Farhan M, Elgarni M (2018) Climate change effects on water resources in Libya. Hydrol Sci J 63(4):512–527
Fekri AM, De Maeyer P (2019) Urban growth monitoring and analysis in the greater Tripoli Region, Libya, using remote sensing and GIS techniques. Remote Sens 11(19):2252
Folland CK, Karl TR, Christy JR, Clarke RA, Gruza G, Jouzel J, Mann ME, Oerlemans J, Salinger MJ, Wang SW (2001) Observed climate variability and change. In: Climate change 2001: the scientific basis (pp. 99–181). Cambridge University Press
Feng X, Fu B, Piao S, Wang S, Ciais P, Zeng Z, ... Wu B (2016) Revegetation in China's Loess Plateau is approaching sustainable water resource limits. Nat Clim Change 6(11):1019–1022. https://doi.org/10.1038/nclimate3053
Frédéric (2020) Groundwater storage changes in the major North African transboundary aquifer systems during the GRACE Era (2003–2016). Water 12(2669):1
Freeman RE (2015) Stakeholder theory. Wiley Encyclopedia of Management 1–6

Gallart F, Llorens P (2003) Catchment management under environmental change: impact of land cover change on water resources. Water Int 28(3):334–340

Gonzales RM (2001) Climate change and water resources in Libya. In: Climate change in the Mediterranean. Springer, pp 197–214

Guendouz A, Moulla AS (2010) The shared resources in the North-Western Sahara Aquifer System (Algeria-Tunisia-Libya): the use of environmental isotopes (Algeria Part). In: International conference "transboundary aquifers: challenges and new directions" (ISARM2010)

GWA (1999) Great Jamahiriya's water situation and the national strategy for water resource management

Hamad. (2012) Status of groundwater resource of Al Jabal Al Akhdar Region, North East Libya. J Environ Hydrol 20(2):11

Hamad M, Ahweej A (2020) Climate cha nge adaptation strategies in North Africa. Environ Stud 11(2):214–230

Hamad M, Saaid M (2022) Assessing climate risk in coastal regions. J Clim Environ 13(1):123–145

Han W, Meehl GA, Rajagopalan B, Fasullo JT, Hu A, Lin J, ... Yeager S (2010) Patterns of Indian Ocean sea-level change in a warming climate. Nat Geosci 3(8):546–550. https://doi.org/10.1038/ngeo901

Hansen J, Ruedy R, Sato M, Lo K (2001) Global warming continues: 2000 was the warmest year on record in the instrumental record. Proc Natl Acad Sci 98(26):14778–14783. https://doi.org/10.1073/pnas.261553698

Hasanean HM (2001) Fluctuations of surface air temperature in the eastern Mediterranean. Theor Appl Climatol 68:75–87

Hibbard K et al (2010) Research priorities in land use and land-cover change for the Earth system and integrated assessment modeling. Int J Climatol 13:2118–2128

Huntingford C et al (2005) Impact of climate forcing on land surface temperatures. Geophys Res Lett 32(14):L14703

Hussin HY, Abdullah NR (2018) Analysis of land use/land cover change using remote sensing and GIS techniques in Tripoli city, Libya. Int J GEOMATE 14(42):37–43

IGRAC and UNESCO (2016) Transboundary aquifers of Africa. Based on: Transboundary Aquifers of the World [Map]

Intergovernmental Panel on Climate Change (IPCC) (1996) Climate change 1995: the science of climate change. Contribution of Working Group I to the Second Assessment Report of the Intergovernmental Panel on Climate Change. Cambridge University Press

Intergovernmental Panel on Climate Change (IPCC) (2001a) Climate change 2001: the scientific basis. Contribution of Working Group I to the Third Assessment Report of the Intergovernmental Panel on Climate Change. Cambridge University Press

IPCC (2005) IPCC special report on carbon dioxide capture and storage. Retrieved from: www.ipcc.ch/publications_and_data/publications_and_data_reports.shtml

Intergovernmental Panel on Climate Change (IPCC) (2007) Climate change 2007: synthesis report. Contribution of Working Groups I, II, and III to the Fourth Assessment Report of the Intergovernmental Panel on Climate Change. IPCC, Geneva

Intergovernmental Panel on Climate Change (IPCC) (2013) Climate change 2013: the physical science basis. Contribution of Working Group I to the Fifth Assessment Report of the Intergovernmental Panel on Climate Change. Cambridge University Press

IPCC (2018) Summary for policymakers. In: Global warming of 1.5°C. An IPCC Special Report

IPCC (2019) IPCC special report on the ocean and cryosphere in a changing climate. Retrieved from https://www.ipcc.ch/srocc/

IPCC (2021a) IPCC working group I contribution to the sixth assessment report. Retrieved from https://www.ipcc.ch/report/ar6/wg1/

IPCC (2021b) IPCC working group II contribution to the sixth assessment report. Retrieved from https://www.ipcc.ch/report/ar6/wg2/

IPCC (2021c) IPCC working group III contribution to the sixth assessment report. Retrieved from https://www.ipcc.ch/report/ar6/wg3/

IPCC (2021d) Summary for policymakers. In: Climate change 2021: the physical science basis. An IPCC special report. Retrieved from https://www.ipcc.ch/report/sixth-assessment-report-working-group-i/
IPCC (2021e) Summary for policymakers. In: Climate change 2021: impacts, adaptation, and vulnerability. An IPCC special report. Retrieved from https://www.ipcc.ch/report/sixth-assessment-report-working-group-ii/
IPCC (2021f) Summary for policymakers. In: Climate change 2021: mitigation of climate change. An IPCC Special Report. Retrieved from https://www.ipcc.ch/report/sixth-assessment-report-working-group-iii/
Intergovernmental Panel on Climate Change (IPCC) (2021g) Climate change 2021: the physical science basis. Contribution of Working Group I to the Sixth Assessment Report of the Intergovernmental Panel on Climate Change. Cambridge University Press
Jamet S, Corfee-Morlot J (2009) Assessing the impacts of climate change: a literature review
Jensen ME, Burman RD, Allen RG (1990) Evapotranspiration and irrigation water requirements. ASCE
Kafle D, Bruins HJ (2009) The impact of climate change on regional hydrological components in the Bagmati River Basin in Nepal. Mt Res Dev 29(3):233–242
Kendall MG (1975) Rank correlation measures. Charles Griffin & Company
Libyan Environmental General Authority (2017). https://www.oneplanetnetwork.org/organisations/libya-environment-general-authority-ega
Libyen Environmental General Authority (2017) National report on the state of the environment in Libya. Retrieved from http://www.epa.gov.ly/
Loaiciga HA, El Kenawy AM, Myint SW (2000) Climate change and implications on water resources in Morocco and Tunisia. J Water Resour Plan Manag 126(6):406–415
Mann HB (1945) Nonparametric tests against trend. Econometrica J Econometric Soc 245–259
National Meteorological Centre Libya Meteorological Project Report (2019) https://www.lnmc.ly/demo/aboutus.php#organization
NPC (2014) National planning council national. Water Security Strategy
Pallas (1980) Water resources of the socialist people's Libyan Arab Jamahiriya. In: The Geology of Libya, pp 539–593
Patz JA et al (2005) Impact of regional climate change on human health. Nature 438(7066):310–317
Peng J, Liu Y, Li T, Wu J (2017) Regional ecosystem health response to rural land use change: a case study in Lijiang City, China. Ecol Ind 72:399–410
Peterson TC et al (1995) Global historical climate network data. Bull Am Meterol Soc 76(10):1943–1966
Raja I (2010) Adaptation to climate change in developing countries. Env Policy J 15(3):125–142
Riebsame WE, Meyer WB, Turner BL (1994) Modeling land use and cover as part of global environmental change. Clim Change 28(1):45–64
Rudel TK (2009) How do people transform landscapes? A sociological perspective on suburban sprawl and tropical deforestation. Am J Sociol 115(1):129–154
Salem O (1992) Libya's future water supply the great manmade river project: a partial solution to Libya's future water supply. Int J Water Resour Dev 8:37–41
Schmidt L, Falk T, Siegmund-Schultze M, Spangenberg JH (2020) The objectives of stakeholder involvement in transdisciplinary research: a conceptual framework for reflective and reflexive practice. Ecol Econ 176:106751
Seddon N, Chausson A, Berry P, Girardin CA, Smith A, Turner B (2020) Understanding the value and limits of nature-based solutions to climate change and other global challenges. Philos Trans R Soc B 375(1794):20190120
Shaki A (2006) Simulation and optimization studies for better groundwater management in The Murzuq Basin, Libya
Shukla PR et al (2019) IPCC special report on climate change and land. Intergovernmental Panel on Climate Change

Solomon S (ed) (2007) Climate change 2007—the physical science basis: working group I contribution to the fourth assessment report of the IPCC (vol. 4). Cambridge University Press

Steffen W, Richardson K, Rockström J, Cornell SE, Fetzer I, Bennett EM, Biggs R, Carpenter SR, de Vries W, de Wit CA, Folke C, Gerten D, Heinke J, Mace GM, Persson LM, Ramanathan V, Reyers B, Sörlin S (2018) Planetary boundaries: guiding human development on a changing planet. Science 347(6223):1259855. https://doi.org/10.1126/science.1259855

Stocker TF, Qin D, Plattner GK, Tignor M, Allen SK, Boschung J, Nauels A, Xia Y, Bex V, Midgley PM (2013) Climate change 2013: the physical science basis. Contribution of Working Group I to the Fifth Assessment Report of the Intergovernmental Panel on Climate Change (IPCC). Cambridge University Press

Swailem FM, Hamza MS, Aly AIM (1984) Isotopic composition of groundwaters from Kufra (Libya) as indicator for groundwater formation. Isotopenpraxis Isotopes Environ Health Stud 20(2):64–68

Tateishi R, Ahn CH (1996) Mapping evapotranspiration using remote sensing data. Remote Sens Environ 55(1):48–56

Tellería JL, Blanco E, Guzmán G, González F, Aguilera PA (2019) Land use/cover changes in the Fezzan region (Libya) and the impacts of foreign direct investment in agriculture. Sustainability 11(21):6037

Thapa P, Dhulikhel N (2019) Observed and perceived climate change analysis in the Terai Region, Nepal. GSJ 7(12):35–43

Trenberth KE, Fasullo JT, Balmaseda MA (2016) Earth's energy imbalance. J Clim 29(2):749–760. https://doi.org/10.1175/JCLI-D-15-0768.1

Trenberth KE, Fasullo JT, Shepherd TG (2015) Attribut ion of climate extreme events. Nat Clim Change 5(8):725–730. https://doi.org/10.1038/nclimate2657

Turner BL (1994) Local faces, global flows: the role of land use and land cover in global environmental change. Land Degrad Dev 5(2):71–78

Turner BL, Meyer WB (1994) Global land-use and land-cover change: an overview. Chang Land Use Land Cover Glob Perspect 4(3)

UNECA (2005) Libya national water development report

UNICEF (2019) Assessment of water supply systems and institutions in Libya

Wheida (2012) The water resources management in Libya. Libyan Agric Res Center J Int 3(3):144–154

Xoplaki E, Gonzalez-Rouco JF, Luterbacher J, Wanner H (2003) Mediterranean summer air temperature variability and its connection to the large-scale atmospheric circulation and SSTs. Clim Dyn 20(8):723–739

Xu CY, Singh VP (2005) Evaluation of water resources under changing climate. Hydrol Process 19(6):1329–1354

Zhou Y, Ren G (2011) Change in extreme temperature event frequency over mainland China, 1961–2008. Clim Res 50(2–3):125–139

Ismail Ageena, Ph.D. is an expert at the Libyan National Meteorological Central, a Lecturer at universities in Libya, and an accomplished professional with extensive experience in environmental science, data analysis, statistical consulting, and research. He has also demonstrated expertise in data analysis, statistical methodologies, and research leadership, with a Ph.D. in Environmental Science from the University of Liverpool and a solid background in various roles, including Founder and Supervisor of Aseel Memorial Research Centre (AMRC), Data Analyst and Report Writer at East Lancashire Chamber of Commerce, Researcher at the International Academic Centre (IAC).

Nagib Froja, Ph.D. is an Assistant Professor at Soil and Water Department Agriculture Faculty, Tripoli University, Tripoli-Libya Bachelor in soil and water specializing in water Sciences in 1987. Master of Science in the same field of expertise in 2001, both from Tripoli University in Libya.

Ph.D. from Heriot-Watt University. School of the Built Environment. Water resources department, Edinburgh UK. He has authored numerous articles and papers focusing on climate variability, statistical analysis, geographical analysis, and agricultural production in Libya. Member of several civil society organizations. Through this, he participated in several voluntary civil works related to raising awareness and rationalizing water consumption.

Mr. Salah Hamad is an Assistant Professor of Water Resources and Geoinformation Science at the Department of Natural Resources, Faculty of Natural Resources and Environmental Science, University of Omer Al-Mukhtar, Al Bayda, Libya. He holds an M.Sc. in Geographic Information Systems and Science from Salzburg University, Salzburg, Austria, in 2008. Mr. Hamad has been actively teaching undergraduate courses and supervising bachelor's degree students' graduation since 2010. His research interests lie in GIS mapping and spatial analysis, with a particular focus on geoinformatics research, geodata documentation, and the application of various GIS and remote sensing techniques in conventional water resources management, engineering, and other related aspects of earth science.

Conclusions and Recommendations to the "Water Resources of Libya: Challenges and Management"

Hamdi A. Zurqani

Abstract Libya faces significant challenges in managing its water resources due to population growth, climate change, and inadequate infrastructure. These factors have contributed to limited water resources availability in the country. This has led to a heavy reliance on groundwater as a primary source of drinking water in Libya. However, groundwater extraction for agricultural and domestic purposes causes several challenges, including over-extraction, contamination, and declining water tables. This chapter highlights the most important conclusions and recommendations from all chapters to provide perspective and guidance for students, teachers, researchers, decision-makers, etc. It also includes summarized information to provide readers with an in-depth understanding of the complexities and potential solutions to ensure appropriate management of water resources in the region. Conclusions into different aspects of water management, covering nearly every part of preservation strategies to preserve water quality and ensure equitable distribution among users.

Keywords North Africa · Climate change · Seawater desalination · Population growth · Groundwater depletion · Water management strategies · Water consumption

1 Introduction

Libya has a long history and complex socio-political issues and has struggled to manage its water resources for years. This concluding chapter of "Water Resources of Libya: Challenges and Management" aims to capture the collective wisdom of experts who have dedicated themselves to understanding and addressing this critical issue. Looking forward, it is essential to comprehend the factors that are shaping the water situation in Libya. Historical water use patterns, population growth, political

H. A. Zurqani (✉)
University of Arkansas Agricultural Experiment Station, Arkansas Forest Resources Center, College of Forestry, Agriculture, and Natural Resources, University of Arkansas, Monticello, AR, USA
e-mail: Hzurqani@uark.edu; Zurqani@uamont.edu

H. A. Zurqani (ed.), *Water Resources of Libya*, Springer Water,
https://doi.org/10.1007/978-3-031-80920-0_10

instability, and climate change have all strained the country's ability to secure its water future (Zurqani et al. 2019). Understanding the intricate web of challenges that Libya faces in managing its water resources is necessary for formulating effective solutions. In this context, this book provides a solid foundation for developing recommendations that are not only practical but also mindful of the unique complexities involved.

Throughout the preceding chapters, we have explored the challenges and complexities of water availability and quality in Libya, examining their origins, utilization, and the profound consequences of their mismanagement. We have witnessed the stark reality of water scarcity in a land where survival hinges upon the equitable distribution of this finite and precious resource. Drawing upon the insights gathered in preceding chapters, this final chapter assembles the puzzle pieces of our knowledge into a cohesive picture, offering conclusions illuminating the current state of water resources in Libya and recommendations that can foster a sustainable and water-secure future for this North African country. This chapter summarizes the key findings, insights, and overarching themes discussed in the preceding chapters. In addition, it explores potential management strategies that can pave the way for a more sustainable water future. It also aims to offer actionable recommendations that align with Libya's socio-political realities while considering the environmental impact in the region.

2 The Main Characteristic Issues of Water Resources in Libya

Libya faces a multifaceted water crisis that includes scarcity of surface water, heavy reliance on and overexploitation of groundwater, seawater intrusion, poor water management practices, limited investment in desalination, unequal water distribution, infrastructure challenges, institutional weaknesses, and low public awareness. Addressing these complex issues will require a comprehensive and coordinated approach involving the development of new water sources, optimization of existing water infrastructure and agricultural practices, implementation of effective water conservation policies, and public education campaigns.

2.1 Scarcity of Surface Water

Libya has very limited surface water resources. It has no permanent rivers or freshwater lakes and receives very little annual rainfall. The country is dominated by desert or semi-desert regions, with more than 90% of its land area classified as such (Zurqani 2021). Libya's dearth of surface water has led the country to rely heavily on groundwater as its primary water source (Guwaeder and Ramakumar 2017; Hamad et al. 2017). Population growth and urbanization have further exacerbated the situation, as the increasing demand for water for domestic and agricultural use places significant strain on Libya's limited water resources (Hamad et al. 2017).

2.2 *Heavy Reliance on Groundwater*

Due to the lack of surface water, Libya depends heavily on groundwater, which accounts for over 98% of its total water consumption. However, the overexploitation and mismanagement of groundwater resources have led to a steady decline in groundwater levels, with many aquifers becoming increasingly saline. The groundwater in Libya is the result of hydrological events that occurred thousands of years ago, and the country is currently utilizing this ancient water for the benefit of its people (El-Geriani et al. 1998).

2.3 *Overexploitation of Groundwater*

The extraction of groundwater exceeds the rate of natural replenishment, leading to a decline in water tables and potential depletion of aquifers. This overreliance on groundwater has resulted in the salinization of many of these water sources, further exacerbating the pressing water scarcity challenges faced by Libya as the demand for water continues to grow (Guwaeder and Ramakumar 2017; Hamad et al. 2017).

2.4 *Seawater Intrusion*

Excessive pumping of groundwater along the coast can cause seawater to infiltrate aquifers, contaminating freshwater supplies. This has been a significant issue in Libya, particularly in coastal areas, where the overexploitation of groundwater has led to seawater intrusion, rendering the water unsuitable for domestic and agricultural use (Zurqani 2021). Despite the growing water scarcity in Libya, the previous government has not made significant investments in alternative water supply technologies, such as desalination, which could provide a valuable additional source of freshwater (Brika 2018).

2.5 Poor Water Management

Leakage in water infrastructure, inefficient irrigation practices, and lack of public awareness about water conservation contribute to water waste. Libya's Environmental Law 15 states that water should be used economically and that technologies must be used to minimize water consumption in all activities, but these measures have not been effectively implemented (Goodland 2008). Addressing Libya's complex water challenges will require a multifaceted approach that combines the development of new water sources, the optimization of existing water infrastructure and agricultural practices, and the implementation of comprehensive water conservation policies and public education campaigns.

2.6 Limited Desalination

Desalination is an option for coastal regions, but Libya hasn't fully utilized this technology as a strategic solution. While the establishment of desalination plants in Libya began in the 1960s, the previous government did not make significant investments or prioritize the expansion of desalination to address the country's growing water needs (Ashour and Ghurbal 2004).

2.7 Unequal Water Distribution

Water resources are not evenly distributed across the country. Coastal areas have better access to water compared to inland regions. Providing water to the population living in the northern parts of the country, where most of the population resides, has been a significant challenge for the Libyan government (El-Geriani et al. 1998).

2.8 Infrastructure Challenges

Aging water infrastructure leads to inefficiencies and water losses. Libya has struggled to maintain and upgrade its water supply and distribution network, leading to significant water losses and challenges in meeting the growing demand for water (Hamad et al. 2017).

2.9 Institutional Weaknesses

Lack of clear water management strategies, weak enforcement of regulations, and limited private sector involvement hinder solutions. The Libyan government has not been proactive in implementing comprehensive water management policies, and there has been a lack of coordination among the various institutions responsible for managing the country's water resources (Goodland 2008).

2.10 Low Public Awareness

A lack of understanding about water conservation practices can lead to inefficient water use. Raising public awareness about the importance of water conservation and the need for sustainable water management practices could play a critical role in addressing Libya's water challenges. Furthermore, Public education campaigns can also help to foster a culture of water stewardship, promote the adoption of water-efficient technologies and practices in households and the agricultural sector, and engage the public in collaborative efforts to conserve and manage the country's limited water resources.

Overall, Libya's water crisis is a complex and multifaceted issue that requires a comprehensive and strategic approach to address the various challenges. Developing new water sources, improving water infrastructure, enhancing water management practices, and increasing public awareness are all critical components of a sustainable solution for Libya's water challenges. Furthermore, the Libyan government should focus on developing alternative water sources, such as desalination, to supplement the country's limited groundwater resources (Brika 2018; Goodland 2008). Investing in the maintenance and upgrade of water infrastructure, implementing effective water management strategies, and strengthening the institutional frameworks responsible for water resources management are also essential steps towards ensuring the long-term sustainability of Libya's water supply (Hamad et al. 2017; El-Geriani et al. 1998).

3 Conclusions

The Food and Agriculture Organization (2022) states that Libya is mostly a desert country with limited surface water resources. Nearly all the country's water needs are met by groundwater, most of which is "fossil" water that was recharged during heavy rains thousands of years ago (Issar 1994). However, there is a growing concern about the sustainability of groundwater resources in Libya due to the growing population and increasing agricultural needs (Wahida and Verhoeven 2007). This causes significant challenges for the future management and conservation of water resources in the country. Overusing groundwater seriously threatens the long-term sustainability of this vital resource (Hamed et al. 2018). The groundwater supply in Libya has been severely over-used by these unexpected human activities and increasing economic growth (Abdu Dayem and Scott 2014). It is evident that the challenges facing Libya's water resources are multifaceted and interconnected, requiring a comprehensive and holistic approach to address them effectively. The reliance on fossil water, coupled with population growth and agricultural demands, poses a significant threat to the sustainability of water resources in Libya. (Bindra et al. 2014) To achieve sustainable water management in Libya, it is essential to explore alternative water sources, such as desalination or water recycling, and implement efficient irrigation techniques (Zurqani et al. 2019).

Furthermore, collaborating with international partners and investing in water storage and distribution infrastructure can also help mitigate the challenges Libya faces in managing its water resources effectively. Overall, it is clear that a concerted effort is needed to ensure the sustainable management of water resources in Libya. In conclusion, addressing the complex water challenges in Libya requires a concerted effort from various stakeholders, including government authorities, civil society organizations, and the international community. By embracing sustainable practices and fostering collaboration, Libya can pave the way for a more secure and resilient water future for its people.

4 Recommendations

Managing water resources is essential for any country due to its arid climate and limited water availability. The adverse effects of a lack of water are likely to get worse without effective management and adequate infrastructure. In light of these challenges, Libya must prioritize developing and implementing sustainable water management practices. This may involve investing in water-efficient agricultural technologies, promoting water conservation measures, and exploring alternative water sources such as desalination and wastewater recycling. Additionally, there is a need for robust governance and regulatory frameworks to ensure responsible water usage and equitable distribution. Here are some recommendations for addressing the water resource challenges in Libya:

- **Integrated Water Resource Management (IWRM)**: Implement a comprehensive IWRM approach that considers the interactions between surface water, groundwater, and other natural resources. This approach will help ensure sustainable and equitable use of water resources.
- **Water Conservation and Efficiency**: Promote water-saving technologies and practices in agriculture, such as drip irrigation and efficient crop selection. Encourage water-efficient practices in urban areas, including the use of low-flow fixtures and the reduction of water wastage.

- **Desalination and Water Treatment**: Invest in desalination technologies to augment water supply, mainly for coastal cities. Develop advanced water treatment methods to make use of brackish groundwater and recycled wastewater for non-potable purposes.
- **Groundwater Management**: Implement measures to control over-extraction of groundwater, including setting sustainable extraction limits and monitoring aquifer levels. Promote artificial recharge techniques to replenish groundwater resources.
- **Rainwater Harvesting and Storage**: Encourage the construction of rainwater harvesting systems for individual households and larger community facilities. Develop storage infrastructure to capture and store rainwater for agricultural and domestic use.
- **Water Pricing and Subsidy Reform**: Gradually reform water pricing to reflect the actual cost of water supply and promote efficient use. Redirect subsidies towards promoting water-saving technologies and practices.
- **Diversification of Water Sources**: This can include desalination plants, water reuse, groundwater, and water transfers. For example, investigate the feasibility of using treated wastewater for industrial processes and irrigation. Diversification reduces the pressure on one source and minimizes the risk of running out of water during times of scarcity.
- **Public Awareness and Education**: Launch campaigns to raise public awareness about the value of water and the need for its efficient use. Educate farmers, urban residents, and industrial sectors about sustainable water management practices.
- **Data Collection and Monitoring**: Establish a comprehensive water information system to monitor water resources, consumption patterns, and water quality. Use collected data to inform decision-making and track progress toward water management goals.
- **Policy and Regulatory Framework**: Develop and enforce water management policies that promote sustainability and equitable distribution of water resources. Collaborate with international organizations and experts to draw upon best practices and experiences.
- **Climate Resilience Planning**: Integrate climate change considerations into water resource management strategies to adapt to changing precipitation patterns and increased evaporation.
- **Capacity Building and Institutional Strengthening**: Invest in training programs for water professionals and government officials to enhance their understanding of modern water management techniques. Strengthen institutions responsible for water resource management and governance.

References

Abdudayem A, Scott AH (2014) Water infrastructure in Libya and the water situation in agriculture in the Jefara region of Libya. Afr J Econ Sustain Develop 3(1):33–64

Ashour MM, Ghurbal SM (2004) Economics of seawater desalination in Libya. Desalination 165:215–218

Bindra SP, Hamid A, Salem H, Hamuda K, Abulifa S (2014) Sustainable integrated water resources management for energy production and food security in Libya. Procedia Technol 12:747–752

Brika B (2018) Water resources and desalination in Libya: a review. In: Proceedings (vol 2, No 11, p 586). MDPI. https://doi.org/10.3390/proceedings2110586

El-Geriani AM, Essamin O, Loucks DP (1998) Water from the desert: minimizing costs of meeting Libya's water demands. Inst Operat Res Manag Sci 28(6):23–35. https://doi.org/10.1287/inte.28.6.23

Food and Agriculture Organization of the United Nations (FAO) (2022) Libya water scarcity and climate change. Food and Agriculture Organization of the United Nations. https://www.unicef.org/mena/media/19321/file/Libya%20water%20scarcity%20analysis%20and%20recommendations_%20UNICEF%20Sep%202022.pdf

Goodland R (2008) How Libya could become environmentally sustainable. Cambridge University Press 39:145–160. https://doi.org/10.1017/s0263718900010049

Guwaeder A, Ramakumar R (2017) A study of grid-connected photovoltaics in the Libyan power system. Energy Power 7:41e49

Hamad JRJ, Hanafiah M, Yaakob WZW (2017) Water resources management in Libya: challenges and future prospects. Zibeline Int Publish 1(2):02–05. https://doi.org/10.26480/mjsa.02.2017.02.05

Hamed Y, Hadji R, Redhaounia B, Zighmi K, Bâali F, El Gayar A (2018) Climate impact on surface and groundwater in North Africa: a global synthesis of findings and recommendations. Euro-Mediterr J Environ Integr 3:1–15

Issar AS (1994) Fossil water under the deserts of the Middle East: a safeguard against strife. In: Studies in Environmental Science (vol 58). Elsevier, pp 351–361

Wheida E, Verhoeven R (2007) An alternative solution of the water shortage problem in Libya. Water Resour Manage 21:961–982

Zurqani HA (ed) (2021) The soils of Libya. World soils book series (1st ed). Springer, Cham. ISBN-13: 978-3030663674

Zurqani HA, Mikhailova EA, Post CJ, Schlautman MA, Elhawej AR (2019) A review of Libyan soil databases for use within an ecosystem services framework. Land 8(5):82

Dr. Hamdi A. Zurqani is an Assistant Professor of Geospatial Science in Natural Resource Management and Conservation at the University of Arkansas Agricultural Experiment Station, Arkansas Forest Resources Center, University of Arkansas at Monticello, USA. With a strong academic foundation and a passion for innovative technologies, he is also an FAA (i.e., Pt107) Certified sUAS/Drones Pilot, leveraging this expertise to enhance his proficiency in remote sensing, GIS, and geospatial data analysis, which he applies to advance his research and teaching in environmental monitoring and analysis. Dr. Zurqani is a recognized expert as a result of his internationally acclaimed work in the areas of environmental information science, remote sensing, geospatial analysis, land evaluation, sustainability, pedology, and soil science education. He has conducted research across the world, including the United States of America and Africa, and has served as PI, co-PI, or co-investigator on several grants-funded research projects. In addition to his research and teaching responsibilities, Dr. Zurqani is actively involved in institutional service at the University of Arkansas at Monticello. He serves on the UAM Campus-wide Artificial Intelligence (AI) Review Committee, the UAM Faculty Assembly Student Engagement Committee, and the UAM Library Committee.

Professionally, Dr. Zurqani contributes to the broader academic community through a variety of leadership and collaborative roles. He is an active member of the Steering Committee of the Institute of Electrical and Electronics Engineers (IEEE) Mediterranean and Middle-East Geoscience and Remote Sensing Symposium (M2GARSS) and serves as an Ambassador for the Geoscience and Remote Sensing Society (GRSS) Chapter of the IEEE in Libya, facilitating communication, collaboration, and knowledge-sharing within the geospatial and remote sensing communities. In recognition of his expertise and substantial contributions to the field, Dr. Zurqani was invited in 2024 to serve on the Technical Program Committee for the IEEE Asia-Pacific Conference on Geoscience, Electronics, and Remote Sensing Technology (AGERS). In 2025, he also joined the IEEE GRSS Technical Committee on Remote Sensing Environment, Analysis, and Climate Technologies (REACT), a platform for scientists and engineers to exchange ideas and knowledge on environmental issues and the impacts of climate change, emphasizing support for sustainable development goals (SDGs) through remote sensing technologies. Beyond these roles, Dr. Zurqani holds the position of UAM Co-Principal Investigator at the Southeast Climate Adaptation Science Center, where he is involved in research initiatives focused on climate adaptation and resilience in the southeastern United States and its surrounding regions.

Dr. Zurqani's editorial and peer-review activities further highlight his commitment to advancing scientific research. He serves on the Editorial Boards of several journals, including *Frontiers in Earth Science* (Geoinformatics), *Discover Soil* (Springer Nature), *Frontiers in Remote Sensing*, and *Remote Sensing* (MDPI). He has also been appointed as a Guest Editor for special issues in several journals, focusing on topics such as advancing spatial prediction of soil properties using remotely sensed data and geospatial artificial intelligence, and UAV agricultural management. Dr. Zurqani's dedication to his field is further demonstrated by his role as a peer reviewer for numerous publishers, including Springer Nature, Nature, Frontiers, Elsevier, and MDPI, among others. He is also the editor of several influential publications that address critical environmental challenges. His notable works include: *The Soils of Libya* (2021), part of Springer's World Soils Book Series; *Environmental Applications of Remote Sensing and GIS in Libya* (2022); and *Water Resources of Libya: Challenges and Management* (2025), published as part of the Springer Water Series (SPWA). These publications highlight his multidisciplinary expertise and commitment to addressing top environmental issues in Libya and beyond.

The manufacturer's authorised representative in the EU is Springer
Nature Customer Service Centre GmbH, Europaplatz 3, 69115 Heidelberg,
Germany. If you have any concerns regarding our products, please
contact ProductSafety@springernature.com

Printed and bound by CPI Group (UK) Ltd, Croydon, CR0 4YY
01/10/2026
02225949-0003